T0282440

Advances in Intelligent Systems and Computing

Volume 426

Series editor

Janusz Kacprzyk, Polish Academy of Sciences, Warsaw, Poland
e-mail: kacprzyk@ibspan.waw.pl

About this Series

The series "Advances in Intelligent Systems and Computing" contains publications on theory, applications, and design methods of Intelligent Systems and Intelligent Computing. Virtually all disciplines such as engineering, natural sciences, computer and information science, ICT, economics, business, e-commerce, environment, healthcare, life science are covered. The list of topics spans all the areas of modern intelligent systems and computing.

The publications within "Advances in Intelligent Systems and Computing" are primarily textbooks and proceedings of important conferences, symposia and congresses. They cover significant recent developments in the field, both of a foundational and applicable character. An important characteristic feature of the series is the short publication time and world-wide distribution. This permits a rapid and broad dissemination of research results.

Advisory Board

Chairman

Nikhil R. Pal, Indian Statistical Institute, Kolkata, India
e-mail: nikhil@isical.ac.in

Members

Rafael Bello, Universidad Central "Marta Abreu" de Las Villas, Santa Clara, Cuba
e-mail: rbellop@uclv.edu.cu

Emilio S. Corchado, University of Salamanca, Salamanca, Spain
e-mail: escorchado@usal.es

Hani Hagras, University of Essex, Colchester, UK
e-mail: hani@essex.ac.uk

László T. Kóczy, Széchenyi István University, Győr, Hungary
e-mail: koczy@sze.hu

Vladik Kreinovich, University of Texas at El Paso, El Paso, USA
e-mail: vladik@utep.edu

Chin-Teng Lin, National Chiao Tung University, Hsinchu, Taiwan
e-mail: ctlin@mail.nctu.edu.tw

Jie Lu, University of Technology, Sydney, Australia
e-mail: Jie.Lu@uts.edu.au

Patricia Melin, Tijuana Institute of Technology, Tijuana, Mexico
e-mail: epmelin@hafsamx.org

Nadia Nedjah, State University of Rio de Janeiro, Rio de Janeiro, Brazil
e-mail: nadia@eng.uerj.br

Ngoc Thanh Nguyen, Wroclaw University of Technology, Wroclaw, Poland
e-mail: Ngoc-Thanh.Nguyen@pwr.edu.pl

Jun Wang, The Chinese University of Hong Kong, Shatin, Hong Kong
e-mail: jwang@mae.cuhk.edu.hk

More information about this series at http://www.springer.com/series/11156

Michel-Alexandre Cardin
Saik Hay Fong · Daniel Krob
Pao Chuen Lui · Yang How Tan
Editors

Complex Systems Design & Management Asia

Smart Nations – Sustaining and Designing:
Proceedings of the Second Asia-Pacific
Conference on Complex Systems Design &
Management, CSD&M Asia 2016

 Springer

Editors
Michel-Alexandre Cardin
Department of Industrial and Systems
    Engineering
National University of Singapore
Singapore
Singapore

Saik Hay Fong
ST Engineering
C/o Singapore Technologies Dynamics
    Pte. Ltd.
Singapore
Singapore

Daniel Krob
CESAMES
Paris
France

Pao Chuen Lui
National Research Foundation
Singapore
Singapore

Yang How Tan
Defence Science & Technology Agency
Singapore
Singapore

ISSN 2194-5357   ISSN 2194-5365   (electronic)
Advances in Intelligent Systems and Computing
ISBN 978-3-319-29642-5   ISBN 978-3-319-29643-2   (eBook)
DOI 10.1007/978-3-319-29643-2

Library of Congress Control Number: 2015961029

© Springer International Publishing Switzerland 2016
This work is subject to copyright. All rights are reserved by the Publisher, whether the whole or part
of the material is concerned, specifically the rights of translation, reprinting, reuse of illustrations,
recitation, broadcasting, reproduction on microfilms or in any other physical way, and transmission
or information storage and retrieval, electronic adaptation, computer software, or by similar or
dissimilar methodology now known or hereafter developed.
The use of general descriptive names, registered names, trademarks, service marks, etc. in this
publication does not imply, even in the absence of a specific statement, that such names are exempt
from the relevant protective laws and regulations and therefore free for general use.
The publisher, the authors and the editors are safe to assume that the advice and information in this
book are believed to be true and accurate at the date of publication. Neither the publisher nor the
authors or the editors give a warranty, express or implied, with respect to the material contained
herein or for any errors or omissions that may have been made.

Printed on acid-free paper

This Springer imprint is published by SpringerNature
The registered company is Springer International Publishing AG Switzerland

Foreword

A System is a set of elements in interaction (Bertalanffy 1968). Advances in information technology have enabled interactions between elements that used to be isolated in space or time. It is this connectivity which increases the complexity in systems. Traditionally, more emphasis is placed on engineering improvements than the design process. However, in recent times there is a growing awareness on the need to combine specialized technical expertise with systems thinking, and to place design at the core to address the increased complexity. Quoting leading systems theorist, Russell Ackoff, "The system loses its essential nature if the individual elements are separated. In some cases, improving the performance of a single component element will actually worsen the performance of the system as a whole." It is thus essential to engage problems from a systems-level perspective.

Many large-scale systems in Singapore were developed through systems engineering. Critical systems such as the transport system, water treatment plants, health care, energy distribution, supply chain management require a systems-level design to achieve optimized performance at world-class standards. Moving forward, we will continue to be challenged by increasingly multifaceted issues brought about by rapid urbanization, global warming, security, an aging society, and a new wave of interconnected revolution (e.g., Internet of things, etc.). To continue to be relevant to the twenty-first century, it is thus important to equip graduates with new mindsets and skills that can provide practical, sustainable solutions cutting across traditional boundaries. SUTD was established in 2009 with Engineering Systems and Design, Information Systems Technology and Design, Engineering Product Development and Architecture and Sustainable Design as its four core pillars. SUTD aims to nurture technically-grounded leaders and innovators to serve societal needs through an integrated multidisciplinary curriculum and multidisciplinary research. We are very excited to be given the opportunity to partner with CESAMES and other sponsors to host CSD&M Asia 2016 on our new campus as we see great synergy between the goal of the conference and the mission of SUTD. With many experts from academia, industry, and the government sector converging at this conference, many great ideas will be generated and collaborations will be forged. It is my hope that this conference will inspire researchers and engineers to take up the challenge of achieving engineering feats which can create a positive and lasting impact for the future of the world.

A warm welcome to Singapore!

Prof. Tow Chong Chong
Provost of Singapore University of Technology
and Design, Director of Temasek Laboratories at SUTD

Preface

Introduction

This volume contains the proceedings of the Second International Asia-Pacific Conference on "Complex System Design & Management" (CSD&M Asia 2016; see the conference website: http://www.2016.csdm-asia.net/ for more details).

The CSD&M Asia 2016 conference was jointly organized during February 24–26, 2016 at the Singapore University of technology and Design by the two following founding partners:

1. The Singapore University of Technology and Design (SUTD);
2. The Center of Excellence on Systems Architecture, Management, Economy and Strategy (CESAMES).[1]

The conference also benefited from the permanent support of other organizations such as the DTSA Academy (Singapore), the National University of Singapore, and many other institutions to which a number of committee members belong, and deeply contributed to the conference organization.

Special thanks also go to Accenture Technology Labs Beijing (China), Dassault Systèmes (France), DSO National Laboratories (Singapore), Defence Science and Technology Agency of Singapore (DSTA—Singapore), Electricité de France (EDF —France/Singapore), International Council on Systems Engineering (INCOSE), IRT SystemX (France), JTC Corporation (Singapore), Mega International (France), Obeo (France), Project Performance International (PPI), Sembcorp (Singapore), Surbana Jurong (Singapore), and Thales (France/Singapore), which were our key industrial and institutional sponsors. The generous support of Sembcorp is especially pointed out here.

---

[1]CESAMES is a nonprofit organization, dedicated to the organization of CSD&M conferences, which was created by the Ecole Polytechnique—Dassault Aviation—DCNS—DGA—Thales "Engineering of Complex Systems" chair.

We are also grateful to the Council of Engineering Systems Universities (CESUN), Dassault Systèmes Singapore, Data Analytics Technologies & Applications Research Institute (Taiwan), French Chamber of Commerce in Singapore, French Embassies in Asia, Gumbooya (Australia), INCOSE Asia-Oceania Sector, Infocomm Development Authority of *Singapore* (IDA), Institution of Engineers Singapore (IES), Land Transport Authority (Singapore), Ministry of Defence of Singapore (MINDEF & SAF), Ministry of Home affairs (Singapore), National Research Foundation (NRF, Singapore), National Instruments South East Asia (Singapore), PUB-The Singapore National Water Agency, Singapore Power, Systematic (France), Veolia Environment (France/Singapore), and Vinci Constructions (France/Singapore), who all strongly supported our communication efforts.

All these institutions also helped us a lot through their constant participation in the organizing committee during the one-year preparation of CSD&M Asia 2016.

Many thanks therefore to all of them.

Why a CSD&M Asia Conference?

Mastering complex systems requires an integrated understanding of industrial practices as well as sophisticated theoretical techniques and tools. This explains the creation of an annual *go-between* forum in the Asia-Pacific region dedicated both to academic researchers and industrial actors working on complex industrial systems architecture, modeling, and engineering. Facilitating their *meeting* was actually for us a *sine qua non* condition in order to nurture and develop in the Asia-Pacific region the new emerging science of systems.

The purpose of the conference on "Complex Systems Design & Management Asia" (CSD&M Asia) is exactly to be such a forum, in order to become, in time, *the* Asia-Pacific academic–government–industrial conference of reference in the field of complex industrial systems architecture and engineering. This is a quite ambitious objective, which we think possible to achieve, based on the success of the "mother" conference, that is to say, the CSD&M conference is ongoing in France since 2010 with a growing audience (the last 2015 edition gathered almost 300 participants coming from 20 different countries with an almost perfect 50/50 balance between academia and industry).

Our Core Academic–Industrial Dimension

To make the CSD&M Asia conference a convergence point for academic, government, and industrial communities interested in complex industrial systems, we based our organization on a principle of *complete parity* between academics, government, and industrialists (see the conference organization sections). This principle was first implemented as follows:

- Program Committee consisted of 50 % academics and 50 % government/industrialists;
- Invited speakers came from numerous professional environments.

The set of activities of the conference followed the same principle. They indeed consist of a mixture of research seminars and experience sharing, academic articles, governmental and industrial presentations, cutting edge software presentations, etc. The conference topics cover in the same way the most recent trends in the emerging field of complex systems sciences and practices from an industrial, governmental, and academic perspective, including the main industrial and public domains (aeronautic and aerospace, defense and security, electronics and robotics, energy and environment, health and welfare services, media and communications, software and e-services, transport, technology & policy), scientific and technical topics (systems fundamentals, systems architecture and engineering, systems metrics and quality, systems modeling tools) and system types (transportation systems, embedded systems, software & information systems, systems of systems, artificial ecosystems).

The Second CSD&M Asia 2016 Edition

The CSD&M Asia 2016 edition received 43 submitted papers, of which the program committee selected 17 regular papers to be published in these proceedings, which corresponds to about a 39 % acceptance ratio which is fundamental for us to guarantee the high quality of the presentations. The program committee also selected 7 papers for a collective presentation during the poster workshop of the conference that intends to encourage presentation and discussion on other interesting, emerging issues.

Each submission was assigned to at least two program committee members, who carefully reviewed the papers, in many cases with the help of external referees. These reviews were discussed by the Program Committee co-chairs during a meeting held at DSTA in Singapore on October 8, 2016, and managed using the EasyChair conference management system. Our sincere thanks go also to Dr. William Yue Khei Lau from DSO National Laboratories, whose help was precious during this evaluation step.

We also invited 15 outstanding speakers from various industrial, governmental, and scientific backgrounds, who gave a series of invited talks covering the entire spectrum of the conference on the theme of "Smart Nations: Sustaining and Designing," mainly during the two first days of CSD&M Asia 2016. The first and second days of the conference were especially organized around this common topic, following up on the discussions from the 2014 edition focusing on "Designing Smart Cities." This theme gave coherence to all invited talks and between the two editions organized so far. The last day was dedicated to special "thematic open sessions," followed by presentations of all accepted papers as well as a system-focused tutorial in parallel.

Furthermore, we had a systems architecture and engineering tools session in order to provide to each participant a good vision the present status of the systems engineering services and tools offers.

Singapore   Michel-Alexandre Cardin
Singapore   Saik Hay Fong
Paris   Daniel Krob
Singapore   Pao Chuen Lui
Singapore   Yang How Tan
November 2016

Acknowledgments

Finally, we would like to thank all members of the Program and Organizing Committees for their time, effort and contributions to make CSD&M Asia 2016 a top quality conference. A special thank is addressed to the CESAMES team (see http://www.cesames.net/), the DSTA Academy team and to the Singapore University of Technology and Design team who managed permanently with a huge efficiency all the administration, logistics and communications of the CSD&M Asia 2016 conference.

The organizers of the conference are also grateful to all the following sponsors and partners without whom CSD&M Asia 2016 would simply not exist:

Founding Partners

- Center of Excellence on Systems Architecture, Management, Economy and Strategy (CESAMES),
- Singapore University of Technology and Design (SUTD).

Academic Sponsors

- Ecole Polytechnique,
- National University of Singapore (NUS).

Professional Sponsors

- Accenture Technology Labs Beijing,
- Dassault Systemes,
- DSO National Laboratories,
- Defence Science and Technology Agency (DSTA),
- Electricité de France (EDF),
- International Council on Systems Engineering (INCOSE),
- IRT SystemX,
- JTC Corporation,
- Mega International,
- Obeo,

- Project Performance International (PPI),
- Sembcorp,
- Surbana Jurong,
- Thales.

Supporting Partners

- Council of Engineering Systems Universities (CESUN),
- Data Analytics Technologies & Applications Research Institute (Taiwan),
- Dassault Systèmes Singapore,
- French Chamber of Commerce in Singapore,
- French Embassies in Asia,
- Gumbooya Pty Ltd.,
- INCOSE Asia-Oceania Sector,
- Infocomm Development Authority of *Singapore* (IDA),
- Institution of Engineers Singapore (IES),
- Land Transport Authority (Singapore),
- Ministry of Defence of Singapore (MINDEF & SAF),
- Ministry of Home affairs (Singapore),
- National Research Foundation (Singapore),
- National Instruments South East Asia (Singapore),
- PUB-Singapore National Water Agency,
- Singapore Power,
- Systematic (France),
- Veolia Environment (France/Singapore),
- Vinci Constructions (France/Singapore).

Paris & Singapore,   Michel-Alexandre Cardin
November 2015   National University of Singapore
                                                                        Singapore

                                                                    Saik Hay Fong
                                                            ST Engineering, Singapore

                                                                        Daniel Krob
                                                CESAMES & Ecole Polytechnique, France

                                                                    Pao Chuen Lui
                                                        National Research Foundation
                                                            Prime Minister's Office
                                                                        Singapore

                                                                    Yang How Tan
                                            Defence Science & Technology Agency Academy
                                                                        Singapore

Organization

Conference Chairs

General Chair
Daniel Krob, Institute Professor, Ecole Polytechnique, France
Pao Chuen Lui, Advisor, National Research Foundation, Prime Minister's Office, Singapore

Organizing Committee Chair
Yang How Tan, President, Defence Science and Technology Agency (DSTA) Academy, Singapore (Chair)

Program Committee Co-chairs
Michel-Alexandre Cardin, National University of Singapore, Singapore (academic co-chair)
Saik Hay Fong, ST Engineering, Singapore (industrial co-chair)
Yang How Tan, DSTA Academy, Singapore (governmental co-chair).

Program Committee

The Program Committee consisted of 28 members (academic, industrial and governmental) who are personalities of high international visibility. Their expertise spectrum covered all of the conference topics. The members of this committee were in charge of reviewing and rating carefully each submission and selecting the best of them for these proceedings.

Co-chairs

- Michel-Alexandre Cardin, National University of Singapore, Singapore (academic co-chair)
- Saik Hay Fong, ST Engineering, Singapore (industrial co-chair)
- Yang How Tan, DSTA Academy, Singapore (governmental co-chair).

Members

- Erik Aslaksen, Gumbooya Pty Ltd., Australia
- Petter Braathen, Memetor, Norway
- Aaron Chia, National University of Singapore, Singapore
- Lynette Cheah, Singapore University of Technology and Design, Singapore
- Stefano Galelli, Singapore University of Technology and Design, Singapore
- Nuno Gil, University of Manchester, United Kingdom
- Paulien Herder, TU Delft, Netherlands
- Joseph Kasser, Temasek Defense Systems Institute at NUS, Singapore
- Serge Landry, Thales Group, Singapore/France
- Erick Lansard, Thales Group, Singapore/France
- William Yue Khei Lau, DSO National Laboratories, Singapore
- Grace Lin, Technologies and Applications Research Institute, Taiwan
- Udo Lindemann, TU Munich, Germany
- Kelvin Ling, Sembcorp, Singapore
- Dong Liu, Accenture Technology Labs Beijing, China
- Xiao Liu, Shanghai Jiao Tong University, China
- Chandran Nair, National Instruments South East Asia, Singapore
- Karthik Natarajan, Singapore University of Technology and Design, Singapore
- William Nuttall, The Open University, United Kingdom
- Jean-Claude Roussel, Airbus Group Innovations, France
- Seiko Shirasaka, Keio University, Japan
- Afreen Siddiqi, MIT, United States of America
- Eun Suk Suh, Seoul National University, Korea
- Fei Yue Wang, Chinese Academy of Science, China
- Erik Wilhelm, Singapore University of Technology and Design, Singapore
- Kristin Wood, Singapore University of Technology and Design, Singapore
- Laura Wynter, IBM Research Collaboratory, Singapore
- Maria Yang, MIT, United States of America.

Organizing Committee

The Organizing Committee consisted of 14 members (academic, industrial, and governmental) in charge of the program and the logistical organization of the conference.

Chair

- Yang How Tan, DSTA Academy, Singapore (chair).

Members

- Samuel Chan, Land Transport Authority, Singapore
- Jimmy Khoo, Singapore Power, Singapore
- Bernard Koh, PUB-The National Water Agency, Singapore
- Erick Lansard, Thales Group, Singapore /France
- William Yue Khei Lau, DSO National Laboratories, Singapore
- Yew Weng Lee, Surbana Technologies Pte. Ltd., Singapore
- Grace Lin, Data Analytics Technologies & Applications Research Institute, Taiwan
- David Long, INCOSE & VITECH, United States of America
- Nicolas Marechal-Abram, Veolia City Modeling Center, Singapore/France
- Chandran Nair, National Instruments South East Asia, Singapore
- Chee Seng Tan, MINDEF & SAF, Singapore
- Su Chern Tan, JTC Corporation, Singapore
- Philippe Vancapernolle, Vinci Construction, Singapore/France
- Pablo Viejo, Electricité De France, Singapore/France.

Invited Speakers

Societal Challenges—Technology & Policy

- Thomas Magnanti, President, Singapore University of Technology and Design, Singapore
- Gee Paw Tan, Chairman, PUB-The National Water Agency, Singapore
- Teng Chye Khoo, Executive Director, Centre for Liveable Cities, Ministry of National Development, Singapore
- Kok Yam Tan, Head, Smart Nation Programme Office (SNPO), Singapore
- Richard de Neufville, Professor, MIT, United States of America

Panel 1 led by Peter HO, Senior Adviser, Centre for the Strategic Futures, Singapore

Industrial Challenges

- Peng Yam Tan, Chief Executive, Defence Science & Technology Agency (DSTA), Singapore

- Erick Lansard, Vice President Technical & Space Development, Thales, Singapore/France
- Soon Poh Heah, Assistant CEO Engineering & Operations Group, JTC Corporation, Singapore.

Panel 2 led by Kerry Lunney, Chief Engineer & Director, Thales/INCOSE of Asia-Oceania Sector—Australia

Scientific State of the Art

- Daniel Krob, Institute Professor, CESAMES & Ecole Polytechnique, France
- Theresa Brown, Distinguished Member of Technical Staff & Program Manager, Sandia National Laboratories, United States of America
- Alan D. Harding, Head of Systems Engineering, Defence Information at BAE Systems/President-Elect, INCOSE, United Kingdom
- See Kiong Ng, Program Director, Urban Systems Initiatives, Agency of Science, Technology and Research (A*STAR), Singapore.

Methodological State of the Art

- Kate Smith-Miles, Professor, Monash University, Australia
- Kristin Wood, Professor, Singapore University of Technology and Design, Singapore
- Jimmy Khoo, Managing Director, Singapore District Cooling at Singapore Power Ltd., Singapore

Tutorial

"Big Data & Data Analytics"

Curated and led by Dr. See Kiong Ng—Program Director, Urban Systems Initiatives, Agency of Science, Technology and Research (A*STAR)—Singapore. Presenting 5 subject-matter experts sharing a wide range of topics:

- Stories that Data Tells: Use Cases of Big Data and Analytics, (Shonali Krishnaswamy, Data Analytics Department, Institute for Infocomm Research, A*STAR)
- Learning from Big Data: New Machine Learning Paradigms, (Shaowei Lin, Sense & Sense-abilities Department, Institute for Infocomm Research, A*STAR)
- Computing on Big Data: Architectures and Algorithms, (Artur Andrzejak, Heidelberg University)

- Developing a Big Data and Analytics Platform for Smart Cities—A*DAX, (Wee Siong Ng, Urban Systems Initiative, Institute for Infocomm Research, A*STAR)
- Robustness in Decision-Making in Uncertainty, contributed by Professor Karthik Natarajan, Singapore University of Technology and Design—Singapore.

Contents

Part II   Posters

Contributors

Sridhar Adepu iTrust, Center for Cyber Security Research, Singapore University of Technology and Design, Singapore, Singapore

Hatim Adli Dassault Systèmes, Vélizy-Villacoublay, France

Helene Bachatene Thales Research and Technology, Palaiseau, France

Mohd. Faisal Bin Zainal Abiden Thales Solutions Asia, Singapore, Singapore

Mihal Brumbulli PragmaDev, Paris, France

Lynette Cheah Engineering Systems and Design, Singapore University of Technology and Design, Singapore, Singapore

Waqas Cheema Engineering Systems and Design, Singapore University of Technology and Design, Singapore, Singapore

Stephen Kheh Chew Chai Thales Solutions Asia, Singapore, Singapore

Eng Seng Chia Temasek Defence Systems Institute, National University of Singapore, Singapore, Singapore

Christopher Choo Engineering Systems and Design, Singapore University of Technology and Design, Singapore, Singapore

Zhongwang Chua Temasek Defence Systems Institute, National University of Singapore, Singapore, Singapore

Matteo Convertino HumNat Lab, Division of Environmental Health Sciences and PH Informatics Program, School of Public Health, Institute on the Environment and Institute for Engineering in Medicine, University of Minnesota, Twin Cities, MN, USA

Daniel Dahlmeier SAP Singapore, Singapore, Singapore

Jitamitra Desai Nanyang Technological University, Singapore, Singapore

Laurence Dooley Faculty of Mathematics, Computing and Technology, The Open University, Walton Hall, MK, UK

Antoine Fagette Thales Solutions Asia, Singapore, Singapore

Christophe Fernandes Dassault Systèmes, Vélizy-Villacoublay, France

Paulo Manuel Ferrão IN+, Instituto Superior Tecnico, Lisbon, Portugal

Jun Wen Fong Temasek Defence Systems Institute, National University of Singapore, Singapore, Singapore

Eliane Fourgeau Dassault Systèmes, Vélizy-Villacoublay, France

Emmanuel Gaudin PragmaDev, Paris, France

Arevik Gevorgyan Laboratoire d'Informatique (LIX), École Polytechnique, Palaiseau, France; Strategy—IP Platforms, Alcatel-Lucent International, Boulogne-Billancourt, France

Emilio Gomez Dassault Systèmes, Vélizy-Villacoublay, France

Wael Hafez Independent Researcher, Alexandria, VA, USA

Michel Hagege Dassault Systèmes, Vélizy-Villacoublay, France

Tan Chuan Heng Land Transport Authority, Singapore, Singapore

Oh Sin Hin Land Transport Authority, Singapore, Singapore

Hao Ming Huang Thales Solutions Asia, Singapore, Singapore

Makoto Ioki Graduate School of System Design and Management, Keio University, Kohoku-Ku, Yokohama, Kanagawa, Japan

Wang Jin SAP Singapore, Singapore, Singapore

Nobuyuki Kobayashi Keio University, Graduate School of System Design and Management, Kohoku-Ku, Yokohama, Kanagawa, Japan

Daniel Krob CESAMES & Ecole Polytechnique, Paris, France

Guoquan Lai Temasek Defence Systems Institute, National University of Singapore, Singapore, Singapore

Kah Wah Lai DSO National Laboratories, Singapore, Singapore

Jean Henri Loic Lancelin Thales Solutions Asia, Singapore, Singapore

Serge Landry Thales Solutions ASIA PTE, LTD, Singapore, Singapore

Zhengyi Lian Defence Science and Technology Agency (DSTA), Singapore, Singapore

Jiwei Lin Institute of Catastrophe Risk Management (ICRM), Interdisciplinary Graduate School, Nanyang Technological University, Singapore, Singapore

Yang Liu HumNat Lab, Division of Environmental Health Sciences and PH Informatics Program, School of Public Health, Twin Cities, USA

Hai Yun Lu SAP Singapore, Singapore, Singapore

Jianxi Luo Singapore University of Technology and Design, Singapore, Singapore

Aditya Mathur iTrust, Center for Cyber Security Research, Singapore University of Technology and Design, Singapore, Singapore

George Matthew Faculty of Mathematics, Computing and Technology, The Open University, Walton Hall, MK, UK

Gérard Memmi CNRS LTCI-UMR 5141, Télécom ParisTech, Paris, France

Ben Mestel Faculty of Mathematics, Computing and Technology, The Open University, Walton Hall, MK, UK

Hui Min Ng Thales Solutions Asia, Singapore, Singapore

Barnabé Monnot Engineering Systems and Design, Singapore University of Technology and Design, Singapore, Singapore

Richard de Neufville Institute for Data, Systems, and Society, Massachusetts Institute of Technology, Cambridge, MA, USA

William J. Nuttall Faculty of Mathematics, Computing and Technology, The Open University, Walton Hall, MK, UK

Yuki Onozuka Graduate School of System Design and Management, Keio University, Kohoku-Ku, Yokohama, Kanagawa, Japan

Mohamad Azman Bin Othman Land Transport Authority, Singapore, Singapore

Georgios Piliouras Engineering Systems and Design, Singapore University of Technology and Design, Singapore, Singapore

Rakesh Prakash Nanyang Technological University, Singapore, Singapore

Marcus Shihong Wu Defence Science and Technology Agency, Singapore, Singapore

Seiko Shirasaka Graduate School of System Design and Management, Keio University, Kohoku-Ku, Yokohama, Kanagawa, Japan; Keio University, Graduate School of System Design and Management, Kohoku-Ku, Yokohama, Kanagawa, Japan

Mong Soon Sim DSO National Laboratories, Singapore, Singapore

Mong Leng Sin DSO National Laboratories, Singapore, Singapore

Peter Spencer Strategy—IP Platforms, Alcatel-Lucent International, Boulogne-Billancourt, France

Yanjun Sun CNRS LTCI-UMR 5141, Télécom ParisTech, Paris, France

Kang Tai School of Mechanical and Aerospace Engineering, Nanyang Technological University, Singapore, Singapore

Sheng Yong Kenny Teo Temasek Defence Systems Institute, National University of Singapore, Singapore, Singapore

Siow Hiang Teo Defence Science and Technology Agency (DSTA), Singapore, Singapore

Shawn Thian Thales Solutions Asia, Singapore, Singapore

Sidney Tio Thales Solutions Asia, Singapore, Singapore

Robert L.K. Tiong School of Civil and Environmental Engineering, Nanyang Technological University, Singapore, Singapore

Keng-Hoe Toh Thales Solutions ASIA PTE, LTD, Singapore, Singapore

Antoine Truong Thales Solutions ASIA PTE, LTD, Singapore, Singapore

Sylvie Vignes CNRS LTCI-UMR 5141, Télécom ParisTech, Paris, France

Erik Wilhelm Engineering Product Development, Singapore University of Technology and Design, Singapore, Singapore

Kevin Williams Temasek Defence Systems Institute, National University of Singapore, Singapore, Singapore

Chee Mun Kelvin Wong Temasek Defence Systems Institute, National University of Singapore, Singapore, Singapore

Nasa Yoshioka Keio University, Graduate School of System Design and Management, Kohoku-Ku, Yokohama, Kanagawa, Japan

Seyed Mehdi Zahraei Engineering Systems and Design, Singapore University of Technology and Design, Singapore, Singapore

Yuren Zhou Engineering Product Development, Singapore University of Technology and Design, Singapore, Singapore

Part I
Invited and Regular Papers

The Need for Technology Policy Programs: Example of Big Data

Richard de Neufville

Abstract The modern mature state is necessarily responsible for designing and managing complex systems. These frequently involve both technological and political issues simultaneously. The modern state therefore requires cadres of professionals cognizant and skilled in technology and policy. Yet many countries, in particular those that have recently transitioned to the state of being modern states, traditionally educate their prospective leaders exclusively in either technical or social expertise. The conclusion is that modern mature states need educational research programs that prepare cadres for leadership in Technology Policy. An example makes the point. As suggested by the case of Singapore, coherent policy formation for "big data", for the collection and use of massive data on individuals and collective enterprises, requires leadership that is both sensitive to the political issues, and knowledgeable about the technological potential.

1 Overall Argument

This presentation argues that modern mature states need to establish and nurture programs in Technology and Policy. The role of these programs would be to prepare future leaders for intelligent, thoughtful, and effective formation and implementation of policies in the wide range of societal issues that necessarily involve advanced technologies in some fashion.

This paper develops this conclusion through logical analysis using syllogisms. A syllogism is the process of proceeding from a major premise A, through a minor premise B, to a conclusion C. For example:

- Major premise: "All human beings are mortal."

R. de Neufville (✉)
Institute for Data, Systems, and Society, Massachusetts Institute of Technology,
77 Massachusetts Avenue, Cambridge, MA 02139, USA
e-mail: ardent@mit.edu

© Springer International Publishing Switzerland 2016
M.-A. Cardin et al. (eds.), *Complex Systems Design & Management Asia*,
Advances in Intelligent Systems and Computing 426,
DOI 10.1007/978-3-319-29643-2_1

3

- Minor premise: "Politicians are human beings."
- Conclusion: "Politicians are mortal."

The strength of such arguments of course depends on the validity of each premise. To make a convincing argument it is thus necessary to establish the credibility of the premises. For example, the conclusion that "Politicians are mortal" rests upon the strength of the claims that "All human beings are mortal" and that "Politicians are human beings."

This presentation uses two syllogisms to make its argument. The first makes the case for the need for leadership cadres who can integrate technological and policy considerations. The second syllogism combines this preceding conclusion with the observation that current educational processes generally do not prepare the requisite cadres, to arrive at the conclusion that mature societies need to fill this gap and create situationally appropriate forms of educational and research programs in Technology Policy.

The paper illustrates the overall argument using the case of "big data." This is the term many observers now commonly use to describe:

> "… the large volume of data – both structured and unstructured – that inundates a business on a day-to-day basis. But it's not the amount of data that's important. It's what organizations do with the data that matters. Big data can be analyzed for insights that lead to better decisions and strategic business moves."
>
> http://www.sas.com/en_us/insights/big-data/what-is-big-data.html?gclid=CN-XtLS6k8kCFRUOjgod4KQMgA#dmhistory

While many thinkers have discussed the issues associated with the torrents of information that now flow into databases, common lore attributes the creation, or at least the popularization, of the "big data" term to industry analyst Douglas Laney.

2 Need for Leadership that Integrates Technology and Policy

The syllogism to develop this part of the argument is:

- Major premise: "States oversee complex technical/social systems.'
- Minor premise: "Proper oversight of these systems benefits from proficiency in technology policy."
- Conclusion: "States benefit from having proficiency in technology policy."

2.1 States Oversee Complex Technical/Social Systems

Mature states necessarily oversee the design and management of complex systems. Their societies expect that they will be able to switch on lights, obtain potable water from a tap, communicate electronically with their friends, manage human services and so on. They expect their governments to make sure that there is the proper organizational structure to fulfill these expectations.

The emphasis here is on overseeing the functioning of these complex systems to serve societal expectations, in contrast to providing for these needs directly. Indeed, mature governments have evolved a wide range of organizational structures to provide expected services to the public. These cover a broad range, for example as regards the provision of water services:

- Direct provision of services through some governmental agency such as the Singapore Public Utilities Board that supplies water nationally;
- Self-sufficient special purpose independent public agencies, such as the East Bay Municipal Utility District (EBMUD) that supplies a range of cities in the San Francisco Area;
- Concession agreements for a limited period, as for the Metropolitan Waterworks and Sewerage System of Greater Manila; and
- Private companies that the government regulates, as through the Water Services Regulation Authority (Ofwat) in England and Wales.

The point is that in every case the government has, as it must, some degree of control over the provision of these complex systems, either directly or indirectly through appointments of the managers, contractual obligations, or regulatory processes.

Typically, the management of the complex systems falls within the domain of either technical professionals or leaders with some formation in the social sciences. For example, engineers typically plan and design the systems that provide water services, electric power, or communication networks. Complementarily, lawyers, politicians, and social scientists generally manage the range of social services the public expects of the state. It is thus easy to think that in the modern state the management of complex systems is split between two cultures: that of science and technology on the one hand, and that of the liberal professions on the other [1].

However, although we might characterize the management of each of these complex systems as either 'technical' or 'social policy', each inevitably combines elements of the other. For example, much of the rationale for the provision of potable water is a matter of social policy concerned with public health and the fear that high rates would encourage the use of contaminated supplies. Conversely, the effective delivery of social services depends strongly on the use of technological means to identify members of the public requiring these services, and to deliver the intended support to them.

The conclusion is that the modern state does oversee complex systems that jointly, to some degree, combine a mixture of technological and social dimensions and concerns. This validates the major premise.

2.2 Oversight Benefits from Proficiency in Technology Policy

Experience demonstrates that the design and management of complex systems frequently benefit from a joint proficiency and understanding of both technology and policy. Example after example shows that lack of understanding of both aspects can lead to difficulties, a waste of time and money, and a generally inefficient, ineffective management of complex systems.

Cases in point that are perhaps most accessible publically concern the way technical managers have been blind to social realities and political processes. The history of the development of the Interstate highway system in the United States is a general example. Engineers in state highway planning agencies used technical models to specify the routes and width of the highways. This approach worked reasonably well for planning motorways through the open countryside, but largely failed in urban areas. The highway planning mindset at that time, in the 1960s, simply did not grasp either the reality that urban populations had different priorities than the traffic engineers, or the need for a negotiated political resolution of conflicting perspectives. They did not know how to deal with citizen groups that were interested in maintaining neighborhoods, in preventing the division of their cityscapes into islands separated by 4 or 6-lane highways. Moreover, the highway agencies demonstrated little understanding of political processes. This lack of appreciation for social and political realities led to extended battles that led to notable costly, time-wasting defeats for highway departments in Boston, San Francisco, and many other cities. They had neither understanding nor skills in developing or implementing effective Technology Policy.

The lack of understanding of the "other", of policy processes by technologists or of technology by liberal professionals, is not confined to the scientific or technical communities. For example, the economists who designed the original privatization of electric power in England set up a system they thought would provide fair competition and keep prices competitive. Unfortunately, they apparently did not understand the way power networks work. In a nutshell, they were ignorant of Kirchhoff's Laws. The result was that they inadvertently created market rules that allowed the major providers of electricity to game the system and essentially extract monopoly prices from consumers. They thus cost the British consumers an enormous amount. Eventually, the British Government revised the market for electric power through a new regulatory system, following which prices reportedly dropped by 40 %—which is a measure of the cost to the public of managing the power system without a good understanding of its technological realities [2].

These two examples document the potential cost of trying to design and manage complex systems with the proper understanding of both the technological and political economic aspects of the situation. This validates the minor premise.

2.3 States Benefit from Proficient Technology Policy

From the preceding we can conclude that modern mature states can benefit from leadership that is knowledgeable in and understanding of both the technological and political economic aspects of the complex issues that they oversee.

The emphasis here must be on the potential benefits. That is, that states can benefit significantly from cadres of leaders who have developed an integrated understanding of both the technological and political economic aspects of an issue. This is not to say that every senior engineer should also be politically savvy, or that every economist or politician should have a significant understanding of the technology that underpins the system of interest. It asserts that they are a number of significant situations in which an understanding of technology and policy may be vital.

3 Need for Technology Policy Education/Research Programs

The syllogism to develop this part of the argument is:

- Major premise: "States benefit from having proficiency in technology policy."
- Minor premise: "Technology Policy programs are generally unavailable."
- Conclusion: "States need to develop Technology Policy programs."

The previous section has already made the case for the major premise.

3.1 Technology Policy Programs Are Generally Unavailable

There are only a few Technology Policy programs that prepare cadres for leadership. Salient programs are:

- Technology Policy Programme at the University of Cambridge (UK)
- Engineering and Public Policy at Carnegie-Mellon University (USA)
- Technology, Policy, and Management at the Technical University of Delft (the Netherlands)
- Science and Technology Policy at George Mason University (USA)
- International Science and Technology Policy at George Washington University (USA)
- Science, Technology, and Public Policy at Harvard University (USA)
- Science, Technology, Engineering, and Public Policy at University College London (UK)

- Technology Policy Program at the Massachusetts Institute of Technology (MIT) (USA)
- Science, Technology, and Environmental Policy at Princeton University (USA)
- Science, Technology, and Public Policy at the Rochester Institute of Technology (USA)
- Science and Technology Policy at Stanford University (USA)
- Institute of Science, Technology, and Policy, Technical University of Zürich (ETHz) (Switzerland)

This is a suggestive list of the major Technology Policy programs worldwide.

We should also note that these programs differ substantially in their focus and concerns. Some are institutionally located in the Schools of Engineering, some in Schools of Management, and others in Schools of Public Policy. Correspondingly, the emphasis of their programs and research activities are quite different. Furthermore some have around 40 years experience (Carnegie-Mellon and MIT), and others are just beginning (University College London and ETHz). In general it is fair to say that there is no standard model for technology policy programs [3].

Most of these educational/research programs are in the United States. Four of these are in Europe, and one of these (ETHz) only started up in 2015. Beyond this, Technology Policy programs hardly exist. This account demonstrates the lack of such programs in general.

It is pertinent to ask at this point: why are Technology Policy programs prevalent in the United States but not elsewhere? It is difficult to answer this question conclusively. But the following explanation is at least suggestive. The fact is that technical education in the United States differs significantly from the patterns that are common elsewhere in the world. Thus the criteria for accrediting engineering programs in the United States includes the requirement for

"... a general education component that complements the technical content of the curriculum and is consistent with the program and institution objectives." [4]

In practice this means that engineering students in the US take a quarter of their credits—that is one full year out of the usual four in North America—studying liberal arts subjects. Notably, this is the case at MIT. The graduates of American engineering and technological programs have received at least a rudimentary grounding in the subjects needed to prepare them for developing leadership in effective design and management of complex systems. The result is that American engineering students are primed to develop a range of skills that integrate technical and political economic concerns.

This approach to technical education contrasts with the traditional practice outside of North America, that holds that engineering students study technical matters almost exclusively. That pattern largely prevails in Europe, in educational systems derived from European traditions, and in many Asian contexts. For example, the Dutch university system has traditionally divided its faculties into quite separate groups: Alpha (for languages and the arts); Beta (for technical subjects); and Gamma (for the social sciences). Putting this bluntly, many educational systems make a point of keeping technological and political economic education quite distinct and separate. This means that the technically oriented professionals from those educational cultures are quite unprepared for dealing with issues of social or economic policy in any post-graduate curriculum and research program. This situation makes it difficult to develop programs that combine Technology and Policy.

In any case, we may observe that the existing situation validates the minor premise that "Technology Policy programs are generally unavailable".

3.2 States Need to Develop Technology Policy Programs

The conclusion from the logical argument presented is that modern, mature states owe it to themselves—in the interest of designing and managing their complex systems—to develop academic programs of education and research in Technology Policy. In many fields they will benefit from having such a program. The states will benefit from having a cadre of professionals who are competent technically, understanding of the social and political economic realities, and skillful in meshing these competencies in the management of the complex systems they deal with.

The practical question is then: how might it be possible to develop the desirable Technology Policy programs? It is not enough to have a good idea, it is essential to know "how to move the furniture around", to have some effective ways to implement ideas. In particular, it is necessary to have the needed intellectual, professional, and financial resources.

Appropriate human capital is a necessary condition for success in developing Technology Policy programs. It will not be possible to develop effective Technology Programs without persons prepared for the field, with relevant experience, and interested in applying their capabilities. Without these ingredients, we can do nothing.

But the availability of human capital is not sufficient. Some kind of catalyst is indispensible to bring the human capital together for the creation a new kind of educational and research program. The development of Technology Policy programs has an interesting history in this regard. Historically, individual entrepreneurial pioneers founded and developed most of the existing programs. For example, Prof. Granger Morgan created and led the Carnegie-Mellon program for some four decades. Similarly, Prof. Henk Sol led the creation of the Delft program. Almost all the existing Technology Policy programs stem from some kind of individual entrepreneurial desire to create such activities. To date, most Technology Policy programs originated through a 'bottoms-up" process of institutionalization. Only a few programs, such as the recently initiated ETHz program in Switzerland, appear to have come from a "top-down" process, as an expression of high-level institutional strategy. By the way, this observation offers another explanation for the prevalence of Technology Policy programs in the United States; its culture generally values and promotes individual entrepreneurship, in academia in particular.

Modern states cannot rely on chance and uncertain individual initiatives to initiate and develop the capabilities they require. To obtain the resources and capabilities they need, they should plan and adopt a suitable strategy of initiation, encouragement, and implementation. The first phase of a successful strategy is likely to build upon some one or more themes that are nationally salient as policy issues. This approach can ride the wave of existing interest, and of the funding in this concern.

Financial resources are of course also necessary to establish Technology Policy programs. However, the sums required are not extraordinary if the basic human capital exists within the existing university context. Existing programs have largely developed by building on existing faculty and facilities already on staff. Mostly they have needed the organizational and individual commitment to reassign positions around new themes in Technology and Policy. Start-up money is primary needed to create initial momentum and stimulate excitement.

Since this conference on Complex Systems Design and Management is being held in Singapore, let us take a moment to think about out Singapore could set up an exciting program in Technology and Policy:

- Let us first observe that Singapore already has a set of faculty who have prepared in world-class programs in this field—several of them are participating in this conference.
- Second, Singapore also has a range of senior faculty with significant experience in policy at the highest levels.
- Third, it has at least two well-established institutions relevant to the area:

  - The Lee Kuan Yew (LKY) School of Public Policy that already serves as the regional hub in its field, attracting future leaders from across Asia and beyond; and
  - The Institute for Engineering Leadership in the faculty of Engineering at the National University of Singapore.

- Fourth, both the LKY School of Public Policy and the Institute for Engineering Leadership have close associations with a number of foreign professors and experts who have been working in technology policy for decades.

In short, Singapore has the central necessary ingredients for providing regional leadership in technology policy.

Given that Singapore has a strong base, both in terms of human capital and institutionally, how might it proceed to establish a Technology Policy endeavor? How might the Government develop and provide the capabilities it will need in this area? How might it proceed to mount suitable educational curricula? What should it do to mount a meaningful research program that could productively inform the nation? How might it develop the ideas and concrete plans for managing and developing Technology Policy to serve the nation and the region?

As a suggestion I propose that Singapore could build upon its current commitment to create a "Smart Nation". At its core, this is a project to build national connectivity, to create an operational system to collect and comprehend data, and thus to provide the basis for devising and providing enhanced services [5]. As such, it represents an extensive investment in a technologically complex system, one that will surely last many years. This project also raises, indeed brings with it, a broad range of important policy issues, for example:

- How will the nation deal with the consequences of "big data", the availability of immense of amounts of detailed data on so many aspects of our lives? We have already seen many of the ways "big data" has been disrupting established institutional arrangements.
- How should the nation adjust its urban development (if at all) to take advantage of the capability of big data? The Lee Kwan Yew Centre for Innovative Cities has already initiated discussions on this [6].
- What policies should the government establish to protect itself, the public, and commercial enterprises from inappropriate, abusive intrusions into privacy?

It seems clear that the Smart Nation project not only provides a platform for thoughtful developments in Technology Policy, but also indeed calls for research and education in what we might call "big data Technology Policy".

4 The Case of "Big Data"

We now turn to the case of "big data" in some detail. The purpose is to illustrate the need for Technology Policy programs through a specific example. The discussion first gives a brief portrait of the nature of "big data", and generally suggests the aspirational goals people have for this field. The argument next describes, with specific reference to Singapore, how the governments and others are currently managing the development of "big data". This points out a range of apparent gaps that reinforce the need "big data Technology Policy".

4.1 Nature of "Big Data"

The development, exploitation, and use of "big data" offer a range of really exciting and challenging opportunities. To many of us, comparisons of the possible impact of "big data" to the effect of the "Industrial Revolution" of two centuries ago seem reasonable. In this regard, the kind of definition of "big data" cited earlier widely understates the prospects. Indeed, that definition focuses on the quantity of data:

> "... the large volume of data – both structured and unstructured – that inundates a business on a day-to-day basis. But it's not the amount of data that's important. It's what organizations do with the data that matters. Big data can be analyzed for insights that lead to better decisions and strategic business moves."
> http://www.sas.com/en_us/insights/big-data/what-is-big-data.html?gclid=CN-XtLS6k8kCFRUOjgod4KQMgA#dmhistory

Many of us believe that 'big data" will, eventually, fundamentally change the nature of our society. Its ramifications have already changed our patterns of living and being just since the turn of the century. Think about the social media and how small local incidents (a terrorist act, a police confrontation) can spread worldwide and mobilize political forces rapidly. Recognize how new enterprises based on "big data" (such as Uber and Airbnb) have already disrupted established industries and overturned pre-existing regulations and conventions. Consider the potential to pinpoint minute hidden relations between seemingly disparate factors (such as connections between individual mutations on genetic codes and forms of cancer), and thus to establish specific treatments and cures.

In the event, the following discussion builds upon the concept of "big data" that includes both the wide range of supporting technologies (such as sensors, communication systems, and mathematical analyses) *and* the application to an array of societal issues. With the creation of the MIT Institute for Data, Systems, and Society (MIT IDSS) in 2015, MIT explicitly subscribes to this perspective. As the research program of the MIT IDSS states:

> "Advances in technologies, including big data, sensors, and communications networks, combined with increased computational capabilities and the ability to process and analyze vast amounts of data, have created the opportunity to holistically, systematically, and scientifically address complex systems that touch upon every aspect of our modern lives."
> http://idss.mit.edu/research/

4.2 Management and Development of "Big Data"

The technological effort to create the basis for "big data" has been, and will continue to be enormous. What we have accomplished to date is miraculous: just consider what a smartphone can do for us—compared to what was at the forefront of technology in the year 2000. This accomplishment has required enormous achievements in the design and use of physical materials (computer chips, fiber optics, etc.) and in mathematics (network analysis and control, compression of data, etc.). These kinds of technical achievements have and are driving the development of "big data".

It is thus normal that technologists and technical institutions have a commanding role in the development of "big data" projects. Thus in Singapore, the Infocomm Development Authority leads the way in the "Smart Nation" project. It is instructive to look at how they present their task.

Infocomm's view of their task for the development of "big data" in Singapore appears in Fig. 1. It projects the image that the project is all about the technology. The first phase is the creation of an underlying communications network, followed by picking up available data and storing and sharing it in some way. Indeed, their label for the work is the development of the "Smart Nation Platform".

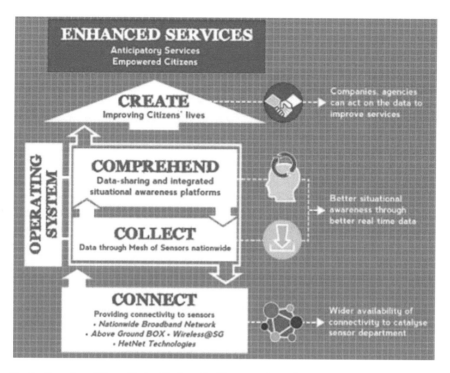

Fig. 1 Overview of Smart Nation Platform for Singapore. https://www.ida.gov.sg/ ∼ /media/Files/
About%20Us/Newsroom/Media%20Releases/2014/0617_smartnation/AnnexA_sn.pdf

Although Fig. 1 identifies the aspirational goals for the Smart Nation Platform as "improving citizens' lives" and "empowering citizens", notice that the immediate task is not confronting the major policy issues involved. Consider the matter of privacy, for example. How is Singapore as a nation going to protect its citizens from unwanted and undesirable intrusion into their intimate details? We can imagine the operational efficiencies that might derive from translating paper medical records into digital formats. In counterpart however, this process facilitates easy access to issues most citizens properly would not want to be publically accessible—such as whether they ever suffered from depression, had a sexually transmitted disease, or underwent an abortion. Unless the state establishes careful policies and practices concerning the privacy of digital personal records, the era of "big data" might be considerably worsen rather than improve the lives of many citizens. In the United States, for example, extended public debates and regulations about health information (US HHS) have taken place. In general, the development of "big data" inherently entails issues of privacy: we do not mind giving various bits of information when we expect it to be lost in a mass of unsearchable data, but we may feel quite differently when we know that governments or companies may access and exploit it. For instance, whereas I now do not mind the obligation to provide my passport or resident ID number when I book a ticket for a theater in Singapore, I well might not want to share the details of my peculiar private viewing pleasures. Thus the Smart Nation project requires careful development of social policies.

We must also recognize that "big data" makes available all kinds of information that was previously unthinkable. The traditional transaction records accessible to governments (land transactions, tax declarations, military and criminal records, etc.) are quite limited compared to what is now possible. Already, the use of smart phones enables the telecom providers—and anyone with whom they wish or must share the data—to track our individual locations minute by minute. Moreover, we can expect the deployment of all kinds of other sensors. Consider the Singapore National Science Experiment (https://www.nse.sg/) that is deploying thousands of wearable sensors (developed in association with the Singapore University of Technology and Design, https://www.nse.sg/sensg/about-sensg/). These enable the researchers to track ambient conditions (temperature, humidity) and activity (sitting or walking) of school children throughout their day. No doubt that this information might help the Land Transport Authority plan its transport services. But which parent wants to broadcast the details of how their child walks home at night? Similarly, while closed circuit TV (CCTV) can be useful in deterring crime or identifying the perpetrators, what should our social policy be concerning its use, in an era when technology will be technically able to tag our faces digitally and our track our activities in real time via CCTV streams?

These social issues are not in evidence on Infocomm's agenda for the Smart Nation Platform. This is proper insofar as their competence and remit centers on technology. The recognition of this reality is simply a statement of fact that identifies the gap between what now exists and the need for Technology Policy programs that somehow jointly address both the technological and the political economic issues associated with the concept of the Smart Nation, and with "big data" generally.

4.3   Need to Develop Technology Policy Programs for "Big Data"

The preceding account stresses the gap between what currently appears to dominate the development of "big data", in Singapore in particular, and what is desirable from the perspective of the state. In a nutshell, the need is for some kind of integrated, joint examination of the technological and the political economic policies that associated with a Smart Nation. Some of this discussion has begun in Singapore [11]. This is a good start. Singapore needs to do more, at a higher level.

Recognizing this kind of need, my colleagues at the MIT Institute for Data, Systems, and Society have made their own commitment in this regard. Specifically, they state their mission in the following terms:

"[Our] research will incorporate both the technological aspects of the application as well as the critical human and institutional aspects inherent in most societal challenges." ...

"IDSS seeks to integrate these areas — fostering new collaborations, introducing new paradigms and abstractions, and utilizing the power of data to address societal challenges." ...

"Our ability to understand data and develop models across complex, interconnected systems is at the core of our ability to uncover new insights and solutions." http://idss.mit.edu/research/

5   Conclusion

The conclusion is that modern mature states need educational programs that prepare cadres for leadership in Technology Policy. Such programs exist at leading universities. These demonstrate what it is possible to achieve. Modern mature states should adopt a coherent strategy to develop such institutions.

Singapore already has the necessary assets, in terms of human capital and institutions, for the development of an effective Technology Policy program. As it is generally most effective to start such programs around salient important issues, the suggestion is that Singapore should use its Smart Nation project as a basis for catalysing its own efforts in Technology Policy.

Such an effort should be valuable to the nation. Coherent policy formation for "big data", for the collection and use of massive data on individuals and collective enterprises, requires leadership that is both sensitive to the political issues, and knowledgeable about the technological potential.

References

1. Snow, C.P.: The Two Cultures and the Scientific Revolution. Cambridge University Press, London (1959/2001)
2. Thomas, S.: Electricity liberalisation: the beginning of the end, Public Services International Research Unit, University of Greenwich Business School (2004)
3. Pond, R.: Liberalisation, privatisation and regulation in the UK electricity sector. http://www.pique.at/reports/pubs/PIQUE_CountryReports_Electricity_UK_November2006.pdf (2006)
4. ABET: Criteria for accrediting engineering programs. http://www.abet.org/wp-content/uploads/2015/05/E001-15-16-EAC-Criteria-03-10-15.pdf (2015)
5. Infocomm Development Authority of Singapore (iDA): Smart Nation Platform. https://www.ida.gov.sg/~/media/Files/About%20Us/Newsroom/Media%20Releases/2014/0617_smartnation/AnnexA_sn.pdf (2014)
6. Lee Kwan Yew Centre for Innovative Cities (LKY CIC), Singapore University of Technology and Design (SUTD): Roundtable of Thought Leaders on Innovation in Cities: Data. Social Behaviour. Policy. http://lkycic.sutd.edu.sg/lkycic-events/roundtable-of-thought-leaders-on-innovation-in-cities-data-social-behaviour-policy-2/ (2015)
7. Institute for Data, Systems, and Society (IDSS), Massachusetts Institute of Technology (MIT). http://idss.mit.edu/about-us/ (2015)

8. US Department of Health and Human Services (US HHS): Health Insurance Portability and Accountability Act (HIPAA): Health Information Privacy. http://www.hhs.gov/ocr/privacy/ (2015)
9. Singapore National Science Experiment, https://www.nse.sg/: What is SENSg? https://www.nse.sg/sensg/about-sensg/ (2015)
10. Institute for Data, Systems, and Society (IDSS), Massachusetts Institute of Technology (MIT). http://idss.mit.edu/research/ (2015)
11. Business Times, Singapore: "Towards a Smart Nation" Six-Part Series (2015)

Towards Model-Driven Simulation of the Internet of Things

Mihal Brumbulli and Emmanuel Gaudin

Abstract The Internet of Things (IoT) refers to the networked interconnection of objects equipped with ubiquitous intelligence, or simply "smart objects". The "smart" part is often followed by words like grid, home, parking, etc., to identify the application domain, and it is provided by software applications and/or services running on top of these large-scale distributed communication infrastructures. Heterogeneity and distribution scale speak for the complexity of such systems and call for a careful analysis prior to any deployment on target environments. In this paper we introduce a model-driven approach for the analysis of IoT applications via simulation. Standard modeling languages, code generation, and network simulation and visualization are combined into an integrated development environment for rapid and automated analysis.

Keywords Modeling · Deployment · Simulation · IoT · SDL · ns-3

1 Introduction

The Internet of Things (IoT) refers to the networked interconnection of billions of "smart objects", i.e., autonomous networked devices equipped with sensors, actuators, computational and storage facilities, that cooperate with each-other and the environment. In recent years IoT has gained much attention from industry and researchers with a variety of applications, e.g., smart-grid [28], smart-home [32], smart-parking [13], earthquake early warning [10], etc. This trend is sure to continue in the immediate future [11], as it continues to be supported by the

M. Brumbulli (✉) · E. Gaudin
PragmaDev, 18 rue des Tournelles, 75004 Paris, France
e-mail: mihal.brumbulli@pragmadev.com

E. Gaudin
e-mail: emmanuel.gaudin@pragmadev.com

© Springer International Publishing Switzerland 2016
M.-A. Cardin et al. (eds.), *Complex Systems Design & Management Asia*,
Advances in Intelligent Systems and Computing 426,
DOI 10.1007/978-3-319-29643-2_2

standardization efforts of well known organizations like ITU-T [20], ETSI [9], IEEE and IETF [16, 17].

Heterogeneity and distribution scale speak for the complexity of IoT applications and call for a careful analysis prior to any deployment on target environments. However, a thorough analysis is possible if the application is already deployed, which contradicts the purpose of the analysis in the first place. A controlled environment (e.g., experimentation test-bed [14]) can be a solution, with the advantage of having the application itself (i.e., the application intended for deployment) under analysis. Nevertheless, scalability of such environments maybe unfeasible, thus hindering the analysis of properties that can be inferred from the interaction of devices deployed in a large-scale. Simulation can address this issue, but the development of an accurate simulation model is not a trivial task. Indeed, a simulation model is an abstraction of the application (i.e., a selection of properties that are relevant to the analysis) and not the application itself. Derivation of this abstraction may become a development process in its own based on the complexity of the system, and validation is a must to ensure that it correctly represents the application intended for deployment.

In Sect. 2 we give an overview of related work on simulation of IoT applications and position our approach in respect to existing state of the art. We introduce our modeling approach by means of a typical application example in Sect. 3. The automatic transformation from model to simulation executable is described in Sect. 4. Finally, we present some conclusions in Sect. 5.

2   Related Work

We identify two types of simulators for IoT applications: specialized and generic. Specialized simulators are provided by IoT operating systems as part of their development environment, e.g., Cooja [27] for Contiki OS [8] and TOSSIM [25] for TinyOS [24]. They simulate the entire device, i.e., hardware, communication, operating system, and application. A major advantage of these simulators is that they do not require a simulation model of the application, i.e., the same description (usually code) can be used for simulation and deployment. However, the number of device models is limited to those supported by the operating system, and scalability is a possible issue as entire devices are simulated. On the other hand, generic simulators (e.g., ns-2 [2], ns-3 [15], and OMNeT++ [34]) can be used to analyze large-scale application-specific properties without the additional overhead of low-level models (e.g., hardware and operating systems), but they require a simulation model of the application.

Weingärtner et al. in [35] discuss possible solutions to the limitations of available simulators. They identify the extension of generic simulators with capabilities of executing real applications or hybrid frameworks as smarter choices.

Although extending generic simulators is quite common, it focuses on models of the underlying communication protocols and not on applications. Tazaki et al. in

[33] describe how to execute nearly unmodified applications in the context of ns-3 simulations. The aim is to increase the number of available protocol models and realism of simulations. The framework is not specifically targeted to IoT applications, however it is an important extension to generic simulation.

Brambilla et al. in [1] present a hybrid framework for the simulation of large-scale IoT scenarios. The framework is characterized by high modularity and loose coupling between simulation models to allow reuse of the code. Although it is possible to deploy the source code on real devices, this can be achieved only by adopting the required interfaces.

We follow a different approach to application development by exploiting the model-driven development (MDD) paradigm. With a pragmatic MDD approach it is possible to construct a model of the application that can be transformed into the application for deployment and/or a simulation model for analysis [31]. We use the standard languages SDL [19] or SDL-RT [30][1] in our integrated development environment[2] to capture architectural, behavioral, communication, and deployment aspects of the application into a single description (model). Automatic code generation can then transform such description into an executable simulation model for the ns-3 network simulator.

3   Modeling

Several efforts have been made in the last decade to bring together standard modeling languages with generic simulation frameworks. The advantages are obvious: while the former allow description of aspects at a higher level of abstraction independent from the target platform, the later are used to simulate such description in large-scale scenarios. Representative examples are the use of SDL with ns-2 [23] or ns-3 [3], and UML with OMNeT++ [7]. There is however an important difference between SDL- and UML-based approaches. While SDL models are used also for deployment (e.g., [10, 22]), this is not the case with UML, where models are used only for simulation.

3.1   Architecture and Behavior

In SDL the overall design is called the *system*, and everything outside of it is defined as the *environment*. The system can be composed of *agents* and *communication constructs*. There are two kinds of agents: *blocks* and *processes*. Blocks can be composed of other agents and communication constructs. When the system

---

[1]A pragmatic combination of the standard languages SDL, UML [26], C/C++.
[2]Real Time Developer Studio—http://www.pragmadev.com.

is decomposed down to the simplest block, the way the block fulfills its functionality is described with processes. A process provides this functionality via extended finite state machines. It has an implicit queue for *signals*. A signal has a name and parameters; they go through *channels* that connect agents and end up in the processes implicit queues. Figure 1 illustrates these concepts in a typical IoT sensor-gateway application example for parking lots.

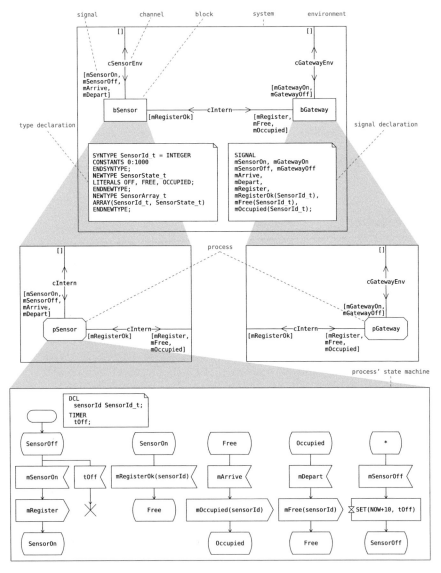

Fig. 1 Architecture and behavior description of a simple sensor-gateway application for parking lots

The gateway keeps an updated list of all sensors it is responsible for. This binding is realized at startup via the signals mRequest and mRequestOk. To keep track of the state of each individual sensor, the gateway assigns a unique identifier to it. At start, every registered sensor is in Free state. When a sensor detects the arrival of a car at its slot (i.e., the mArrive signal), it will change the state to Occupied and let the gateway know about such change by means of the mOccupied signal. This signal has the sensor's unique identifier as parameter, so that it is possible for the gateway to correctly update its list of available slots. When the car leaves the parking slot (i.e., the mDepart signal), the sensor will reset its state to Free and notify the gateway with the mFree signal. Additional signals may trigger the gateway to report the current status of the parking lot, however for simplicity these signals are not shown in the figure.

3.2 Communication

The strong point of SDL is that it allows the description of communication systems in a formal and abstract way (i.e., platform independent). However, this level of abstraction presents a number of challenges when implementation is concerned (i.e., platform specific). In this context, an important aspect to be considered is communication, which puts the internet into the internet-of-things. Communication in SDL is realized using channels, e.g., cIntern in Fig. 1, and it can be local or distributed. Not making the difference between these two types of communication is an important feature of SDL that is needed to abstract from the platform. Indeed, there is no way to tell whether the pSensor and pGateway processes in Fig. 1 are running on the same node or device (i.e., local communication) or on different nodes (i.e., distributed communication). Platform specific implementation for local communication can be derived in general without much effort. The information contained in the SDL architecture and behavior description is enough to uniquely identify the sender and receiver of a signal, because all process instances are part of the same executable running on a single node. Problems begin to emerge when processes are distributed and use platform specific inter-process communication. To uniquely identify a process instance in a distributed IoT infrastructure, information about the node where the instance has been deployed is required; a typical example would be the IP address of the IoT node. However, this information is not present in a SDL model, due also to the missing concept of the node in the language.

Schaible and Gotzhein in [29] define a set of SDL patterns for describing distributed communication. The advantage of these patterns is that they are formalized, thus they can be used in every SDL description without affecting this important feature of the language. However, the introduction of patterns implies changes to an existing SDL description, and what is more important, it does not follow the choice of SDL to abstract the type of communication.

Brumbulli and Fischer in [5] apply the same concept but in the context of SDL-RT. Although not formalized,[3] the patterns are very compact, descriptive, easy to apply, and exploit the pragmatic nature of the language. Nevertheless, the problems of a pattern-based approach are still present, and it is not possible to use the patterns in SDL.

To address these issues we decided to not define and/or apply any pattern to SDL descriptions. The additional information that is required to derive platform specific implementation can be provided using SDL-RT deployment diagrams [30]. This approach does not introduce any changes to the SDL model of a system, thus keeping the desired level of abstraction in the description of communication. Also, by keeping deployment aspects (e.g., nodes and their IP addresses) separate from architecture and behavior, it is possible to use the approach for both SDL and SDL-RT. Furthermore, extending our integrated development environment with the presented approach is an important step towards a complete model-driven solution for the development of IoT applications.

3.3 Deployment

The deployment diagram describes the physical configuration of run-time processing elements of a distributed system and may contain (Fig. 2):

- *nodes* are physical objects that represent processing resources,
- *components* represent distributable pieces of implementation of a system (i.e., SDL block or process),
- *files* provide information about external signals required for simulation.

The node and component are identified by means of the id attribute. The type of values for this attribute depends on the parameters required for inter-process communication, e.g., IP address for the node and TCP port for the component. A comma separated list of values can be assigned to the attribute. This allows the description of large-scale scenarios while keeping the readability of the diagrams. The pair of id attributes is used to uniquely identify each component. The semantics of communication between two process instances are as follows:

- if the sender and receiver instances belong to the same component, then SDL communication semantics apply;
- if the sender and receiver instances belong to different components, then:
  - at first, the pair of identifiers is used to send the signal to the peer component via inter-process communication;
  - afterwards, the signal is delivered to the receiver inside the component using SDL semantics.

---

[3]This does not change anything in the language, because SDL-RT is not formal.

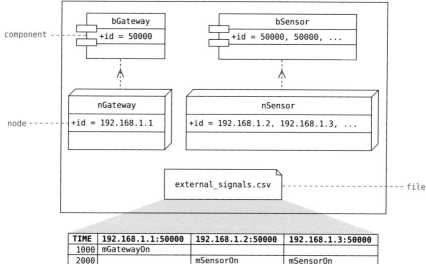

Fig. 2 Deployment scenario for the sensor-gateway application for parking lots

The use of deployment diagrams was introduced [5]. We improve the approach in two significant ways:

- introduce deployment information *exclusively* in the deployment diagram, without mixing architecture, behavior, and deployment aspects of the system, thus supporting both SDL-RT and standard SDL;
- provide simple means to describe interaction with the environment.

Interaction with the environment is an important aspect that must be considered during simulation. An example of such interaction is the detection of arrival and departure of a car into a parking slot. This is modeled with the **mArrive** and **mDepart** signals in Fig. 1. These signals are addressed to the sensor, and it makes perfect sense when the system has a single sensor. However, this is not the case in our example scenario (or every IoT application scenario in general), which is composed of several sensors. In this context, external signals may be addressed to one or more distributed process instances, and to add to the complexity, they can have different parameters, timing, and order of arrival based on the intended receiver. We define a simple yet complete method for extending the deployment model with the required information. The set of external signals with parameters, timing, and receiver is described in a tabular form (i.e., comma separated values format) in a file symbol attached to the deployment diagram as shown in Fig. 2. Each signal is configured with the time it is sent by the environment (row head) and the receiving component (pair of identifiers in column head).

4   Simulation

The SDL (or SDL-RT) description (architecture and behavior) combined with the deployment diagram are used as a basis for the generation of an executable simulation model for ns-3. A generic model for code generation was introduced in [5]. We extend the model as shown in Fig. 3 by introducing the implementation of deployment and distributed communication concepts, i.e., **RTDS_DeplNode**, **RTDS_DeplComponent**, and **RTDS_Proxy**.

The **RTDS_DeplNode** maintains the list of all its attached components in its attribute named componentsList. The **RTDS_DeplComponent** keeps a reference to the scheduler (**RTDS_Scheduler**), which manages process instances (**RTDS_Proc**), communication via signals (**RTDS_MessageHeader**), and timers (**RTDS_TimerState**). Local communication between process instances is handled via a memory shared mechanism. This is possible because local communication implies sender and receiver instances managed by the same scheduler. The information about the sender and receiver of a signal is encapsulated in

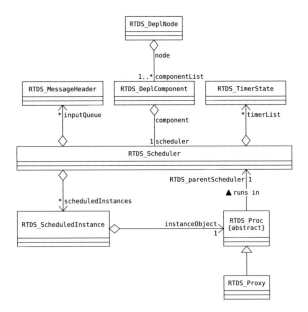

Fig. 3 Extended class diagram for code generation based on [5]

RTDS_MessageHeader. If the sender and receiver are not managed by same scheduler, then distributed communication is implied, and the signal is forwarded to the **RTDS_Proxy instance**. Every scheduler manages an implicit proxy instance, which interfaces to the underlying communication models provided by ns-3.

Our integrated development environment (RTDS) will automatically:

- check the syntax/semantics of the SDL description,
- check the syntax/semantics of the deployment description,
- check the syntax/semantics of the external signals given in tabular form,
- produce an executable by generating the code, compiling it, and linking it to the ns-3 library, and
- launch the executable in the background, interpret and visualize simulation events traced during execution.

Visualization of simulation events is realized as described in [4], but extended with two modes:

- in *live* mode events are visualized as they are traced from the simulation running in the background,
- in *post-mortem* mode the simulation can be replayed entirely after it has successfully terminated, which allows stepping through the events without having to re-run the simulation.

Figure 4 shows the deployment of one gateway and 99 sensors (**nGateway** and **nSensor** in Fig. 2). Nodes, their state, and distributed communication (between nodes) are visualized in the *RTDS Deployment Simulator* as shown in Fig. 4a. Nodes are displayed as colored rectangles, where the color represents the current state of the node. A sent signal is a directed arrow from the sender to the receiver, and its color represents the type of the signal. The arrow is removed from the view when the signal is received. The states and signals can be configured via the provided interface. It is possible to change their color and choose whether to display them during simulation. This feature is useful in cases where not all the states and/or signals are relevant to the analysis.

If an issue is detected in the behavior of the application (e.g., a missed state change or a signal not sent), it is possible to analyze the cause in detail by visualizing the internals of each node. This can be done on-demand using standard MSCs [18] as shown in Fig. 4b, c. MSCs are linked to the SDL description, thus identifying the source of misbehavior in a MSC trace would allow navigation to the corresponding part in the description where changes can be made.

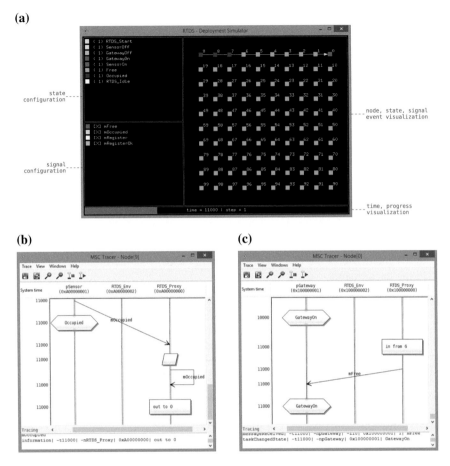

Fig. 4 Visualization of simulation events for the sensor-gateway example

5 Conclusions

In recent years IoT has gained much attention from industry and researchers, a trend that will continue in the immediate future. Heterogeneity and distribution scale contribute to the complexity of IoT applications, which need a careful analysis prior to any deployment.

In this paper we introduced a model-driven approach for the analysis of IoT applications via simulation. The standard language SDL captures architectural, behavioral, and communication aspects into a model of the application that is independent from the target platform. Deployment diagrams describe the distribution of the SDL system (blocks and/or processes) in an IoT infrastructure and the interaction with the environment. The automatic code generation transforms the description into an executable simulation model for the ns-3 network simulator.

The *RTDS Deployment Simulator* provides a graphical interface for the visualization of simulation traces.

The concepts and tools presented in this paper are important steps towards the model-driven analysis of IoT applications via simulation. In this context, the use of standard MSCs for representing traced events can aid the formal verification of properties [12]. As to how this can be applied for applications deployed in large-scale is yet to be investigated in future work. Furthermore, we are considering the possibility to extend the approach with testing support by means of standard TTCN-3 [21]. A deployment model analysis based on symbolic resolution tools such as [6] can be the next step towards a complete and fully automated approach.

References

1. Brambilla, G., Picone, M., Cirani, S., Amoretti, M., Zanichelli, F.: A Simulation Platform for Large-scale Internet of Things Scenarios in Urban Environments. In: Kawsar, F., Blanke, U., Mashhadi, A.J., Altakrouri, B. (eds.) The First International Conference on IoT in Urban Space. pp. 50–55. Urb-IoT '14, ICST (2014)
2. Breslau, L., Estrin, D., Fall, K.R., Floyd, S., Heidemann, J.S., Helmy, A., Huang, P., McCanne, S., Varadhan, K., Xu, Y., Yu, H.: Advances in network simulation. IEEE Comput. **33**(5), 59–67 (2000)
3. Brumbulli, M., Fischer, J.: SDL Code Generation for Network Simulators. In: Kraemer, F., Herrmann, P. (eds.) System Analysis and Modeling: About Models. Lecture Notes in Computer Science, vol. 6598, pp. 144–155. Springer, Berlin/ Heidelberg (2011)
4. Brumbulli, M., Fischer, J.: Simulation Visualization of Distributed Communication Systems. In: Rose, O., Uhrmacher, A.M. (eds.) Proceedings of the 2012 Winter Simulation Conferenc, pp. 248:1–248:12. WSC '12, IEEE (2012)
5. Brumbulli, M., Fischer, J.: Simulation Configuration Modeling of Distributed Communication Systems. In: Haugen, Ø., Reed, R., Gotzhein, R. (eds.) System Analysis and Modeling: Theory and Practice. Lecture Notes in Computer Science, vol. 7744, pp. 198–211. Springer, Berlin Heidelberg (2013)
6. Deltour, J., Faivre, A., Gaudin, E., Lapitre, A.: Model-Based Testing: An Approach with SDL/RTDS and DIVERSITY. In: Amyot, D., Fonseca i Casas, P., Mussbacher, G. (eds.) System Analysis and Modeling: Models and Reusability, Lecture Notes in Computer Science, vol. 8769, pp. 198–206. Springer International Publishing (2014)
7. Dietrich, I., Dressler, F., Schmitt, V., German, R.: Syntony: Network Protocol Simulation Based on Standard-Conform UML 2 Models. In: Glynn, P. (ed.) Proceedings of the 2nd International Conference on Performance Evaluation Methodologies and Tools. pp. 21:1–21:11. ValueTools '07, ICST, Brussels, Belgium (2007)
8. Dunkels, A., Grönvall, B., Voigt, T.: Contiki: A Lightweight and Flexible Operating System for Tiny Networked Sensors. In: Proceedings of 29th Annual IEEE International Conference on Local Computer Networks. pp. 455–462. LCN '04. IEEE Computer Society (2004)
9. ETSI: Machine-to-Machine communications (M2M); Functional architecture. ETSI Technical Specification TS 102 690, European Telecommunications Standards Institute. http://www.etsi.org/deliver/etsi_ts/102600_102699/102690/02.01.01_60 (2013)
10. Fischer, J., Redlich, J.P., Zschau, J., Milkereit, C., Picozzi, M., Fleming, K., Brumbulli, M., Lichtblau, B., Eveslage, I.: A wireless mesh sensing network for early warning. J. Netw. Comput. Appl. **35**(2), 538–547 (2012)
11. Gartner Inc.: Gartner says the Internet of Things installed base will grow to 26 billion units by 2020. http://www.gartner.com/newsroom/id/2636073 (2013)

12. Gaudin, E., Brunel, E.: Property Verification with MSC. In: Khendek, F., Toeroe, M., Gherbi, A., Reed, R. (eds.) SDL 2013: Model-Driven Dependability Engineering. Lecture Notes in Computer Science, vol. 7916, pp. 19–35. Springer, Berlin Heidelberg (2013)
13. Geng, Y., Cassandras, C.: New "smart parking" system based on resource allocation and reservations. IEEE Trans. Intell. Transp. Syst. **14**(3), 1129–1139 (2013)
14. Gluhak, A., Krco, S., Nati, M., Pfisterer, D., Mitton, N., Razafindralambo, T.: A survey on facilities for experimental internet of things research. IEEE Commun. Mag. **49**(11), 58–67 (2011)
15. Henderson, T.R., Roy, S., Floyd, S., Riley, G.F.: ns-3 Project Goals. In: Proceedings of the 2006 Workshop on ns-2: The IP Network Simulator. p. 9. WNS2 '06. ACM, New York (2006)
16. IEEE: IEEE Standard for Local and metropolitan area networks—Part 15.4: Low-Rate Wireless Personal Area Networks (LR-WPANs). IEEE Standard 802.15.4, Institute of Electrical and Electronics Engineers. http://standards.ieee.org/findstds/standard/802.15.4-2011.html (2011)
17. IETF: Compression Format for IPv6 Datagrams over IEEE 802.15.4-Based Networks. Standards Track RFC 6282, Internet Engineering Task Force. http://www.rfc-editor.org/info/rfc6282 (2011)
18. ITU-T: Message Sequence Chart (MSC). ITU-T Recommendation Z.120, International Telecommunication Union—Telecommunication Standardization Sector. http://www.itu.int/rec/T-REC-Z.120/en (2002)
19. ITU-T: Specification and Description Language—Overview of SDL-2010. ITU-T Recommendation Z.100, International Telecommunication Union—Telecommunication Standardization Sector. http://www.itu.int/rec/T-REC-Z.100/en (2011)
20. ITU-T: Overview of the Internet of Things. ITU-T Recommendation Y.2060, International Telecommunication Union—Telecommunication Standardization Sector. http://handle.itu.int/11.1002/1000/11559 (2012)
21. ITU-T: Testing and Test Control Notation version 3: TTCN-3 core language. ITU-T Recommendation Z.161, International Telecommunication Union—Telecommunication Standardization Sector. http://www.itu.int/rec/T-REC-Z.161/en (2014)
22. Kuhn, T., Gotzhein, R., Webel, C.: Model-Driven Development with SDL: Process, Tools, and Experiences. In: Nierstrasz, O., Whittle, J., Harel, D., Reggio, G. (eds.) Model Driven Engineering Languages and Systems. Lecture Notes in Computer Science, vol. 4199, pp. 83–97. Springer, Berlin Heidelberg (2006)
23. Kuhn, T., Geraldy, A., Gotzhein, R., Rothländer, F.: ns + SDL: The Network Simulator for SDL Systems. In: Prinz, A., Reed, R., Reed, J. (eds.) SDL 2005: Model Driven. Lecture Notes in Computer Science, vol. 3530, pp. 1166–1170. Springer, Berli (2005)
24. Levis, P., Madden, S., Polastre, J., Szewczyk, R., Whitehouse, K., Woo, A., Gay, D., Hill, J., Welsh, M., Brewer, E., Culler, D.: TinyOS: An Operating System for Sensor Networks. In: Weber, W., Rabaey, J., Aarts, E. (eds.) Ambient Intelligence, pp. 115–148. Springer, Berlin (2005)
25. Levis, P., Lee, N., Welsh, M., Culler, D.: TOSSIM: Accurate and Scalable Simulation of Entire TinyOS Applications. In: Akyildiz, I.F., Estrin, D., Culler, D.E., Srivastava, M.B. (eds.) Proceedings of the 1st International Conference on Embedded Networked Sensor Systems. pp. 126–137. SenSys '03. ACM, New York (2003)
26. OMG: OMG Unified Modeling Language (OMG UML), Superstructure. Version 2.4.1. OMG Standard, Object Management Group (2011)
27. Österlind, F., Dunkels, A., Eriksson, J., Finne, N., Voigt, T.: Cross-Level Sensor Network Simulation with COOJA. In: Proceedings of the 31st IEEE Conference on Local Computer Networks. pp. 641–648. LCN '06. IEEE Computer Society (2006)
28. Ramchurn, S.D., Vytelingum, P., Rogers, A., Jennings, N.R.: Putting the 'smarts' into the smart grid: a grand challenge for artificial intelligence. Commun. ACM **55**(4), 86–97 (2012)

29. Schaible, P., Gotzhein, R.: Development of Distributed Systems with SDL by Means of Formalized APIs. In: Reed, R., Reed, J. (eds.) SDL 2003: System Design. Lecture Notes in Computer Science, vol. 2708, pp. 158–158. Springer, Berlin (2003)
30. SDL-RT Consortium: Specification and Description Language—Real Time. SDL-RT Standard V2.3, SDL-RT Consortium. http://www.sdl-rt.org/standard/V2.3/html/index.htm (2013)
31. Selic, B.: The pragmatics of model-driven development. IEEE Softw. **20**(5), 19–25 (2003)
32. Silva, L.C.D., Morikawa, C., Petra, I.M.: State of the art of smart homes. Eng. Appl. Artif. Intell. **25**(7), 1313–1321 (2012)
33. Tazaki, H., Urbani, F., Mancini, E., Lacage, M., Câmara, D., Turletti, T., Dabbous, W.: Direct Code Execution: Revisiting Library OS Architecture for Reproducible Network Experiments. In: Almeroth, K.C., Mathy, L., Papagiannaki, K., Misra, V. (eds.) Proceedings of the 9th ACM Conference on Emerging Networking Experiments and Technologies. pp. 217–228. CoNEXT '13. ACM (2013)
34. Varga, A., Hornig, R.: An Overview of the OMNeT ++ Simulation Environment. In: Molnár, S., Heath, J.R., Dalle, O., Wainer, G.A. (eds.) Proceedings of the 1st International Conference on Simulation Tools and Techniques for Communications, Networks and Systems. p. 60. SimuTools '08, ICST, Brussels, Belgium (2008)
35. Weingärtner, E., Ceriotti, M., Wehrle, K.: How to simulate the internet of things? In: Proceedings of the 11th GI/ITG KuVS Fachgespräch Drahtlose Sensornetze. pp. 27–28. FGSN '12. Technische Universität Darmstadt (2012)

Framework for Managing System-of-Systems Ilities

Zhengyi Lian and Siow Hiang Teo

Abstract The design of ilities in System-of-Systems (SoS) architecture is a key means to manage changes and uncertainties over the long life cycle of an SoS. While there is broad consensus on the importance of ilities, there is generally a lack of agreement on what they mean and a lack of clarity on how they can be engineered. This article presents the DSTA Framework for Managing SoS Ilities, which coherently relates key ilities identified as important for SoS architectural design. Newly established in 2013 and updated in 2015 to guide Systems Architecting practitioners in DSTA, the framework also proposes how working definitions of robustness and resilience can be interpreted across key high-level and low-level ilities coherently, and introduces broad concepts of how they could be realized.

Keywords System-of-systems ilities · Robustness · Resilience · Evolvability · Flexibility

1 Introduction

A System-of-Systems (SoS) can be described as a set of constituent systems or elements that are operationally independent, but working together to provide capabilities that are greater than the sum of its parts. Managed independently and distributed geographically, these constituent systems work together to perform some high-level mission which cannot be accomplished by any individual constituent system alone [1]. Figure 1 shows an example of an SoS [2] in which a range

Z. Lian (✉) · S.H. Teo
Defence Science and Technology Agency (DSTA), 1 Depot Road, Singapore 109679,
Singapore
e-mail: lzhengyi@dsta.gov.sg

S.H. Teo
e-mail: tsiowhia@dsta.gov.sg

© Springer International Publishing Switzerland 2016
M.-A. Cardin et al. (eds.), *Complex Systems Design & Management Asia*,
Advances in Intelligent Systems and Computing 426,
DOI 10.1007/978-3-319-29643-2_3

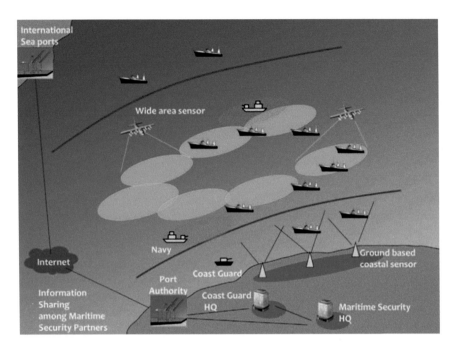

Fig. 1 Example of a maritime security SoS [2]

of systems like coastal and airborne sensors, vessels and information systems from various stakeholders are networked to perform a Maritime Security mission.

Ilities is a term used to describe desirable, lifecycle attributes (usually but not always ending in "ility") of SoS/systems that are not primary functional requirements but manifest themselves after the SoS/system had been put to initial use [3]. Some of such attributes are robustness, resilience, survivability, evolvability, flexibility, adaptability, interoperability, sustainability, reliability, availability, maintainability and safety. For example, robustness and resilience are especially important non-functional attributes for military systems as they are expected to continue performing under harsh environments (even when under attack) and to recover damaged constituents/elements quickly after an attack.

Robustness broadly means maintaining acceptable *performance* across variation in *context*. *Performance* can be interpreted at different levels, such as (a) high-level mission measure of effectiveness (MOE) in MITRE's concept of mission assurance [4], (b) lower level component system measure of performance (MOP) in the United States Department of Defence (U.S. DoD) concept of robust design [5]. Similarly, *context* can be interpreted broadly as mission, operating environment, associated operational contingencies (e.g. deliberate attack, internal failures), constraints, etc. Resilience, in accordance with INCOSE [6], encompasses the ability of a system to absorb (e.g. minimal or graceful degradation) the impact of a disruption or attack, stay at or recover to above an acceptable level of performance and sustain that level for an acceptable period of time.

2 Value and Hierarchy of Ilities

Ilities characterize a system's ability to respond to disturbances and changes in its environment, both foreseeable and unforeseeable. The design of ilities in SoS architecture is a key strategy to achieve consistency in value delivery and to manage changes and uncertainties over the long life cycle of an SoS. The goal is an enduring architecture that continues to perform well under various situations; yet remains flexible and forward-looking to evolve (e.g. allow insertion of new systems, adoption of new concepts of operation) according to changing threats, technology or stakeholder needs.

Despite the well-acknowledged value of ilities, there remains a lack of consensus on what they mean precisely and a lack of clarity of how they can be engineered, except in established areas like quality and RAMS (i.e. Reliability, Availability, Maintainability and Safety). A team comprising de Weck, Ross and Rhodes, from the Massachusetts Institute of Technology Engineering Systems Division (MIT ESD)—Systems Engineering Advancement Research Initiative, conducted a study [7] on 20 common ilities and their inter-relationships (based on co-occurrence in literature[1]), which suggests that a means-end hierarchy of ilities could exist, with certain ilities supporting other ilities. However, a universally accepted hierarchy has yet to be established.

It is desired to create a practicable framework, in the context of the Singapore Armed Forces (SAF) and DSTA, which identifies the key ilities and their relationships to guide the design of ilities in systems architectures. This is the motivation behind the DSTA Framework for Managing SoS Ilities.

3 High-Level (SoS Key Mission) Ilities

We interpret the desired goal for SoS design as Robustness in fulfilling some key high-level mission (i.e. high-level Robustness) under all circumstances. However, practical constraints limit this goal to the mission spectrum and associated operational contingencies defined under a set of foreseeable, well-defined baseline requirements spanning multiple years (see Robustness column of Table 1).

---

[1]Scientific papers within the Inspec and Compendex database (1884–2010) were searched to rank the prevalence of the 20 ilities and found to be consistent with that based on Google search engine hits. Thereafter, an internet search of how often two ilities co-occurred in the same article/page was used to estimate the strength of their inter-relationship.

Table 1 High-level ilities: robustness (in key mission outcome) and evolvability [2]

High-Level Ilities	**Robustness** Can maintain req'd mission effectiveness across mission spectrum and operational contingencies in baseline requirements. Enabled, ultimately, by component systems robustness and resilience.		**Evolvability** Can keep incorporating req'd design changes for high-level mission outcome to remain robust under new requirements arising over time. Enabled by maintaining design flexibility over time.	
Key SoS Reqmts	Baseline Requirements (Foreseeable, Well defined)		New Requirements (Unforeseen, Deferred)	
	Mission Spectrum	Operational Contingencies	New Context (and associated Operational Contingencies)	
Parameter Space for Reqmts	a) Mission Contexts b) Operating Environment c) Threat Types d) Scale e) Manpower Constraints	External (e.g. Disruptive Actions) Internal (e.g. System Failure)	a) Future Threat/Vulnerability b) Future Mission/Tasks c) Future Opportunity d) Future Scale e) Future Constraints	Associated External and Internal Operational Contingencies

Mission spectrum can be described in more detail by parameters such as context, area of operations (e.g. extent and nature), potential threat types, scale (e.g. quantity, level of coordination) and manpower (or other strategic resource) constraints. Operational contingencies can be classified as originating (i) externally, such as disruptive actions (e.g. jamming, cyber-attack, physical damage) by an adversary, extreme natural phenomena (e.g. lightning strikes, heavy storms, earthquakes), or (ii) internally, such as equipment failure, accidents, logistic demand spikes.

High-level Robustness can be measured as the percentage of this baseline requirements parameter space where the high-level mission measure of effectiveness (MOE) stays above some required threshold. In reality, this is challenging due to the many attributes and scenarios to be evaluated in the parameter space. A practical approach is to rationalize and categorize the scenarios by impact and likelihood, followed by further decision on what to include in baseline requirements for high-level Robustness evaluation and what to defer.

The planning space of baseline requirements for high-level Robustness spans multiple years (see Fig. 2). This multi-year planning space can change over the SoS life-cycle to accommodate context changes and new/revised requirements (unforeseen or consciously deferred at the design stage) arising in future.

For example, uncertainty over the materialization of a threat and/or a lack of cost-effective technological solutions may lead to the deferment of requirements to a later stage. These requirements may be re-incorporated when there is more clarity in the threat or solution space. The new requirements will merge with the original baseline requirements to form the newly evolved baseline requirements for the SoS. Some aspects of the original baseline requirements that have become irrelevant over time will be removed or revised in the process.

As another example, abrupt shifts in the Maritime Security landscape (e.g. rise of non-state actors engaging in piracy) may necessitate a significantly higher tempo of operations and/or more surveillance nodes to be established for a Maritime Security

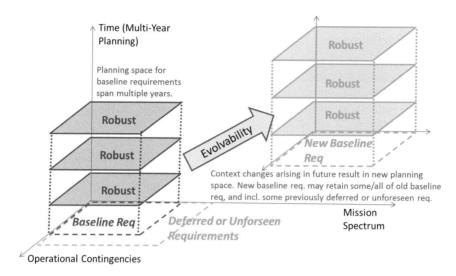

Fig. 2 Planning for high-level robustness spans multiple years. Evolvability is key to maintaining that robustness over time as baseline requirements change

SoS. These changes may create permanent stress to the existing SoS infrastructure (e.g. communications architecture or logistics chain) in a way that necessitates changes to the SoS design. The availability of new technologies (e.g. commoditization of satellite imagery, maturation of fully autonomous patrol vessels) may also offer opportunities for force transformation. These possible shifts in context are likely to translate to revised requirements for the SoS.

Evolvability will allow greater ease to keep incorporating relevant design changes for the SoS to remain robust in fulfilling the high-level mission under each new planning space as new requirements arise over time (see Fig. 2 and Evolvability column of Table 1). This may be measured by the cumulative cost of transition, in terms of capital costs, engineering complexity and/or manpower, associated with anticipated design changes over time, for the purpose of relative comparison between SoS alternatives. In view of the long time horizon and uncertainty of future requirements, such measurements can only be limited to foreseeable new requirements over specified time frames (i.e. the known unknowns).

Hence, Robustness and Evolvability are the two key high-level ilities that an SoS architecture should possess to meet baseline requirements of some key high-level mission, while maintaining design flexibility to meet new/revised requirements (with design change implications) arising over time.

4 Lower-Level (Component Systems Mission or Performance) Ilities

The Robustness of an SoS in fulfilling its key high-level mission (i.e. high-level Robustness) depends on the robustness and resilience of its component systems in fulfilling their lower level missions, which can ultimately be traced to attaining required levels for a set of measures of performance (MOPs) (see Table 2 for examples). Those MOPs should ideally be measurable over time to monitor their levels during normal operations, as well as their degradation (extent and duration) in the event of an operational contingency or disturbance (see Fig. 3).

At the component systems level, robustness can be interpreted as attaining required levels for those MOPs across the SoS mission spectrum and maintaining them under associated operational contingencies (see graphic (A) in Fig. 4). The objective here is for a sufficient subset, *not all*, of the component systems to remain robust under each contingency, so that Robustness in the high level SoS key mission is preserved (see Fig. 5 for concept illustration). In order to sustain this across successive contingencies, component systems resilience is vital to rebuilding capacity to absorb damage from the next wave of contingencies, by limiting damage and recovering affected MOPs of component systems that are more severely impacted by and failed to remain robust under the current wave of contingencies (see graphic (B) in Fig. 4).

There may be cases of contingencies so severe or drastic in impact that existing resilience mechanisms can only recover affected component systems partially. Lessons learnt from such setbacks could trigger a redesign with the aim of recovering affected MOPs to normal/better levels and maintaining them above

Table 2 Examples of SoS/high-level mission and associated lower level component systems, missions and MOPs

SoS and high level mission	Air-power generation	Maritime security
High-level MOE	Probability and time taken to generate req'd level of air-power an attack	Probability and time taken to detect and respond to maritime incidents and threats
Component systems	Individual air-bases or APG clusters	Coastal and sensors, patrol vessels, information systems
Lower level missions	Air-craft generation air-base operability	Surveillance patrol and boarding operations
Possible MOps at a lower level can be related to	Number of aircrafts on standby Aircrafts preparation time Logistics resupply time L/R platform operability L/R platform inspection, recovery key assets and resource availability	Coverage (persistent) of coastal sensors Coverage (transient) revisit time of air-borne sensors and patrol vessels Time to reach incident/threat location Time for completion of boarding operations key assets and resource availability

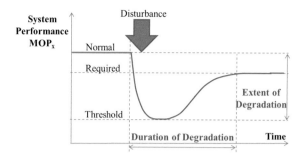

Fig. 3 Disturbances can degrade specific component system MOPs. It is desired for MOP to recover to normal (or at least required) levels quickly after the disturbance, but that may be impossible if the extent of degradation has exceeded some threshold (referenced and adapted from concepts on survivability from MIT ESD [8, 9] and resilience from the Future Resilient Systems project at the Singapore–IEH Centre [10])

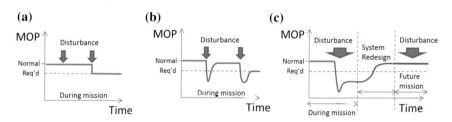

Fig. 4 Component systems responses to operational contingencies: **a** maintain MOP, **b** limit extent and duration of degradation to MOP, **c** design change

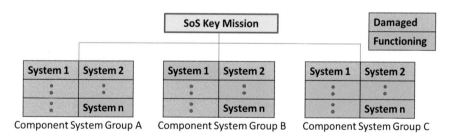

Fig. 5 High-level (SoS key mission) robustness can be preserved after an operational contingency as long as a sufficient subset of component systems (in *green*) remain robust (i.e. relevant MOPs can still attain levels required for mission success)

required levels should the same severe contingencies strike again in future (see graphic (C) in Fig. 4). The flexibility which this design change can be achieved (i.e. design flexibility) without compromising continuity of current operations contributes to SoS evolvability.

Ideally, it would be good to anticipate such severe contingencies and evaluate their impact, either through simulations or exercises, so that the necessary design changes/flexibility can be planned for and implemented progressively at the right opportunity. Possible design choices to preserve design flexibility over time may include (i) availability of redundant component systems to experiment with new designs while preserving capability to perform SoS key missions, (ii) modular task force design to allow easy replacement of platforms or mission modules for different tasks, (iii) avoiding tight coupling, as well as use of closed, proprietary data standards and communication protocols, among component systems.

Table 3 summarizes the above discussion on lower-level ilities from the perspective of component systems responses to operational contingencies. In short, component systems robustness (by attaining required levels for a set of MOPs across the mission spectrum and maintaining them under contingencies) and resilience (by limiting the extent and duration of degradation to those MOPs levels under contingencies) contributes to high-level (SoS key mission) Robustness. Flexibility for design change without compromising continuity in current operations contributes to SoS Evolvability.

5 Achieving Component Systems Robustness and Resilience

Table 3 also represents an attempt to distil out the key design principles, possible broad strategies and enablers that contribute to component systems robustness and resilience, as a guide to generation of solution alternatives for implementation.

Table 3 Component systems responses to operational contingencies: design principles, possible broad strategies and enablers

Lower-Level Ilities	Attain MOP across Mission Spectrum	Component Systems Responses to Operational Contingencies			
		(A) Maintain MOP	(B) Limit extent & duration of degradation to MOP		(C) Design Change
Design Principles at **MOP** Level		← Robustness : Resilience →			To recover MOP (to normal/better) and maintain it if contingency recur in future.
		Avoid Taking a Hit	Avoid/Reduce Damage when Hit	Rapid Recovery ← [Full / Partial] →	
Possible Broad Strategy	Mission & SoS/System Specific	Mobility Concealment Active Def. Diversion	Margin, Hardening	Sense & Repair	Reconfigure, Reorganize ↑↓ Sense	Stress testing & Cap Dev planning to anticipate req'd changes in design or CONOPs and build in req'd flexibility over time.
			Redundancy & Dispersion			
			Sense & Anticipate			
Enablers	"Health" Monitoring (relevant set of MOPs)			Operational Flexibility	Design Flexibility
	RAMS, INTEROPERABILITY, SUPPORTABILITY, LEARNING (what to monitor, how to respond)				

The attainment of required levels for a set of MOPs across a baseline mission spectrum under normal conditions, usually via some mission and SoS/system specific design features, is a typical pre-requisite to be verified during system acceptance tests. However, the above is insufficient to assure component systems robustness and resilience under operational contingencies. That would require additional design features to respond to contingencies along the principles of (1) avoid taking a "hit", (2) avoid/reduce damage when "hit", and (3) rapid recovery.[2] As illustrated in Table 3, principles (1) and (3) contribute to component systems robustness and resilience respectively, while principle (2) contributes to both.[3] These are supported by the enablers "health" monitoring and operational flexibility as shown in Table 3. "Health" monitoring here refers to sensing systems to (i) verify relevant MOPs can attain and maintain required levels over time, (ii) monitor extent of MOP degradation during contingencies for deciding when and where to activate recovery mechanisms, (iii) quantify time taken for recovery. Operational flexibility here encompasses the (i) agility of commanders and operators to quickly respond to contingencies in unfamiliar or untested situations, and (ii) their adaptability to modify responses while sensing overall effectiveness in recovering from the contingency.

5.1 Principle (1)—Avoid Taking a "Hit"

To avoid relevant MOPs being degraded by a disturbance from an external/internal source (i.e. "hit"), possible broad strategies include:

(a) **Mobility**. To stay away from the area of influence of that disturbance.
(b) **Concealment**. To not be easily distinguishable from the environment and be targeted by the source of that disturbance.
(c) **Active Defence**. An active layer to eliminate that disturbance itself or its source, or at least prevent the disturbance from penetrating the boundaries of a system being targeted.
(d) **Diversion**. To deflect/redirect a disturbance that has penetrated a system's boundaries away from critical components of the system.

---

[2]The three principles here are an explicit description of the concepts underlying MIT ESD's survivability design principles of (1) reduce susceptibility, (2) reduce vulnerability and (3) enhance resilience [8, 9].

[3]At the implementation level, we see robustness and resilience as complimentary and overlapping. Robustness measures (avoid a hit, reduce damage when hit) can maintain MOP above required levels for disturbances up to a certain severity. Beyond that, resilience takes over, whereby the same measures to "reduce damage when hit" under robustness now acts to limit damage so that recovery is possible, followed by quick recovery within the mission time-frame. Resilience operates at the tactical level in response to imminent disturbances; evolvability operates at the strategic level in preparation for new/deferred requirements.

5.2   Principle (2)—Avoid/Reducing Damage When "Hit"

When the above layer fails to avoid a component system being "hit" by a disturbance, possible broad strategies to avoid or limit the damage caused include:

(a) **Margin, Hardening**. Building in sufficient margin between required and normal MOP levels allows for some degradation in MOP before the high-level mission effectiveness is compromised. This can be complemented with "hardening" of critical components to resist or slow down the rate of MOP degradation, which can help to lower the amount of margin required.

(b) **Redundancy and Dispersion**. This refers to duplication and avoiding co-location of critical assets/capabilities/operations to avoid creating single-point vulnerabilities/failures. If dispersion via geographical separation is impractical, containment through appropriate barriers to reduce failure propagation (e.g. network separation for IT systems) can be an alternative strategy.

(c) **Sensing and Anticipation**. This applies more to the people or organizations operating an SoS, making sense of anomalies and warning indicators picked up through existing intelligence, sensors or monitoring systems to anticipate an imminent disturbance or detect a developing situation before the damage done becomes too serious. The advance warning can allow operators to respond appropriately (e.g. take cover and avoid exposure, activate barriers or further hardening, getting ready for activation of contingency plans) to reduce potential damage to a minimum. However, a challenge to such a strategy is false alarm and the associated costs. Hence, a fundamental enabler for such a resilience capability is organization learning from past and on-going experience on what to sense/monitor as well as how to respond appropriately to each early warning indicator. (Adapted from Erik Hollnagel's [11] resilient systems/organizations of the third kind.[4])

5.3   Principle (3)—Rapid Recovery

When damage is inevitable after a disturbance, the presence of recovery mechanisms is important to restore affected MOPs quickly and rebuild the capacity to withstand subsequent disturbances. Two possible categories of strategies are:

(a) **Sense and Repair**. "Sense" here refers to using "health" monitoring systems to detect degradation in relevant MOPs, followed by root-cause identification and damage location. Next, decision support tools can help in directing appropriate repair resources to the damage location in the most efficient manner. If the

---

[4]Resilient systems/organizations of the third kind can anticipate and manage something before it happens, instead of passively responding to something that happens.

severity of damage can be limited, this category of response can typically achieve a full recovery to affected MOPs. (Adapted from Erik Hollnagel's [11] resilient systems of the first kind.[5])

(b) **Reconfigure/Reorganize ←—→ Sense**. When damage from a disturbance is so severe that only a partial recovery is possible with existing repair resources, the next strategy is to reconfigure whatever functional capabilities/assets remaining, or to reorganize existing resources, to focus on priority tasks for fulfilling the high-level mission. Contingency planning can help to achieve a quick first response here. However, as the circumstances associated with each contingency evolves differently, a key enabler to deal with such complexity is the operational flexibility of commanders and operators on the ground to regularly "sense" how well the reconfiguration/reorganization is performing and adapt through iterative adjustments to arrive at something that works. Passing down lessons learnt from such episodes can help to sharpen this resilience capability through either better contingency planning for a better first response, or knowing what signals to monitor in making timely adjustments. (Adapted from Erik Hollnagel's [11] resilient systems of the second kind.[6])

6 Fundamental Enablers

The following enablers are fundamental to sustaining effective operations of an SoS over its long life-cycle, which is a pre-requisite to achieving high-level Robustness under baseline (current) and new (future) requirements.

(a) **RAMS**. To operate in an effective and safe manner continuously during the mission period, as well as to instil confidence in sustained mission-readiness over the SoS life-cycle.

(b) **Interoperability**. To enable constituent systems or elements to provide and/or accept data, information, material and services from one another to work coherently in achieving a desired operational effect or high-level mission.

(c) **Supportability**. To sustain effective operations for component systems robustness and resilience over the SoS life-cycle as efficiently as possible, in light of specific resource (e.g. manpower) constraints, through appropriate SoS design (e.g. workflow, technology) at the front end or subsequent SoS evolution.

(d) **Organization Learning**. To accumulate and pass down experience on (i) what to monitor to verify component systems robustness, anticipate potential contingencies, assess need for and effectiveness of responses during contingencies;

---

[5]Resilient systems/organizations of the first kind can monitor situations to determine if a reaction is necessary, and thereafter respond appropriately.

[6]Resilient systems/organizations of the second kind can manage (i.e. monitor and respond to) something not only when it happens, but also learn from what has happened thereafter to adjust both what it monitors and how it responds.

Table 4 DSTA framework for managing SoS ilities

	Robustness		Evolvability	
High-Level Ilities	Can maintain req'd mission effectiveness across mission spectrum and operational contingencies in baseline requirements. Enabled, ultimately, by component systems robustness and resilience.		Can keep incorporating req'd design changes for high-level mission outcome to remain robust under new requirements arising over time. Enabled by maintaining design flexibility over time.	
Key SoS Reqmts	Baseline Requirements (Foreseeable, Well defined)		New Requirements (Unforeseen, Deferred)	
	Mission Spectrum	Operational Contingencies	New Context (and associated Operational Contingencies)	
Parameter Space for Reqmts	a) Mission Contexts b) Operating Environment c) Threat Types d) Scale e) Manpower Constraints	External (e.g. Disruptive Actions) / Internal (e.g. System Failure)	a) Future Threat/Vulnerability b) Future Mission/Tasks c) Future Opportunity d) Future Scale e) Future Constraints	Associated External and Internal Operational Contingencies

Lower-Level Ilities	Attain MOP across Mission Spectrum	Component Systems Responses to Operational Contingencies			
		(A) Maintain MOP	**(B) Limit extent & duration of degradation to MOP**		**(C) Design Change**
Design Principles at **MOP** Level	← Robustness ∣ Resilience →				To recover MOP (to normal/better) and maintain it if contingency recur in future.
Possible Broad Strategy	Mission & SoS/System Specific	**Avoid Taking a Hit:** Mobility Concealment Active Def. Diversion	Margin, Hardening / Redundancy & Dispersion / Sense & Anticipate (**Avoid/Reduce Damage when Hit**)	**Rapid Recovery** ← [Full / Partial] → : Sense & Repair / Reconfigure, Reorganize ↑↓ Sense	Stress testing & Cap Dev planning to anticipate req'd changes in design or CONOPs and build in req'd flexibility over time.
Enablers	"Health" Monitoring (relevant set of MOPs)			Operational Flexibility	Design Flexibility
	RAMS, INTEROPERABILITY, SUPPORTABILITY, LEARNING (what to monitor, how to respond)				

(ii) how to respond (repair, reconfigure, reorganize) and adapt to changing circumstances during contingencies. This will contribute to component systems resilience, and ultimately high-level Robustness.

7   Conclusion

The design of ilities in SoS architecture is a key means to manage changes and uncertainties in environment, technology and stakeholder needs over the long SoS life cycle. The DSTA Framework for Managing SoS Ilities (summarized in Table 4) pioneers the effort to identify and coherently relate key high-level and low-level ilities for an enduring SoS architectural design. In summary, Robustness and Evolvability to are the two key high-level ilities that an SoS architecture should possess. At a lower level, component systems robustness and resilience contribute to high-level Robustness; flexibility for design change without compromising continuity in existing missions contributes to Evolvability. Component systems robustness and resilience to disturbances can be enhanced through strategies along the principles of (1) avoid taking a "hit", (2) avoid/reduce damage when "hit" and (3) quick recovery, supported by SoS "health" monitoring and operational flexibility. Fundamental enablers to sustain effective operations over the SoS life-cycle include RAMS, interoperability, supportability and organization learning.

Acknowledgements The authors would like to thank Mr. Kang Shian Chin and Mr. Sim Kok Wah for initial development of this framework published in DSTA Horizons 2013/14 [2], as well as the perspectives and inputs of Mr. Kang Shian Chin, Mr. Pang Chung Khiang, Mr. Pore Ghee Lye, Dr. Pee Eng Yau and Mr. Wong Ka-Yoon, in the recent effort to update this framework.

References

1. Pang, C.K., Sim, K.W., Koh, H.S.: Evolutionary Development of System of Systems through Systems Architecting. In: Tan Y.H. (ed.) DSTA Horizons 2012, pp. 90–103. DSTA, Singapore (2012)
2. Kang, S.C., Pee, E.Y., Sim, K.W., Pang, C.K.: Framework for Managing System-of-Systems Ilities. In: Tan Y.H. (ed.) DSTA Horizons 2013/14, pp. 56–65. DSTA, Singapore (2013)
3. de Weck, O., Roos, D., Magee, C.: Engineering Systems: Meeting Human Needs in a Complex Technological World, pp. 65–96. MIT Press, Cambridge (2012)
4. Systems Engineering for Mission Assurance. In: MITRE Systems Engineering Guide, pp. 155–157. The MITRE Corporation (2014). http://www.mitre.org/publications/systems-engineering-guide/enterprise-engineering/systems-engineering-for-mission-assurance
5. U.S. Department of Defence, Defence Acquisition University (DAU): Robust Design. In: Glossary of Defence Acquisition Acronyms and Terms, 15th edn. DAU Press, Virginia (2012). https://dap.dau.mil/glossary/pages/2603.aspx
6. INCOSE Resilient Systems Working Group: Working Definition of Resilience. http://www.incose.org/docs/default-source/wgcharters/resilient-systems.pdf?sfvrsn=6 (2011)
7. de Weck, O.L., Ross, A.M., Rhodes, D.H.: Investigating relationships and semantic sets amongst system lifecycle properties (ilities). In: 3rd International Engineering Systems Symposium, CESUN, TU Delft, Netherlands (2012)
8. Richards M.G., Hastings D.E., Rhodes D.H., Ross A.M., Weigel A.L.: Design for survivability: concept generation and evaluation in dynamic tradespace exploration. In: 2nd International Symposium on Engineering Systems, CESUN. MIT, Cambridge (2009)
9. Mekdeci, B., Ross, A.M., Rhodes, D.H., Hastings, D.E.: Examining Survivability of Systems of Systems. In: 2011 INCOSE International Symposium, pp. 569–581. Wiley, Denver (2011)
10. Heinimann, H.R.: Future Resilient Systems. Presentation Slides, 3rd Cities Roundtable, Singapore. www.clc.gov.sg/documents/books/Future%20Resilient%20Systems.pdf (2014)
11. Hollnagel, E.: Resilience Engineering. http://erikhollnagel.com/ideas/resilience-engineering.html (2015)

Expansionism-Based Design and System of Systems

Jianxi Luo

Abstract Facing the growing complexity within technological products and systems, traditional reductionism-based design approaches, which focus on decomposing and optimizing subsystems, components, and their interrelationships, will face greater difficulties in the search for future innovation. In such cases, expansionism-based design will be particularly effective because it reduces the need to deal with the superior internal complexity of existing systems and primarily explores design opportunities by integrating and synthesizing previously unrelated independent systems into a new system of systems. The system of systems that results from expansionism-based design may improve the functionalities and performances of the prior systems, or obtain novel functionalities of the system of systems from the synthesis. In this paper, we identify the system theory roots of design expansionism, and elaborate the value of expansionism-based design and how it enables design opportunities for systems-of-systems. We also preliminarily discuss potential concept generation methods to aid in expansionism-based design and the analytics of collective dynamics and emergent behaviors of the resulting system of systems in order to effectively architect and manage them.

1 Introduction

The technological products and systems that we design and use today are growing in complexity, with the inclusion of more and more components and parts and their increasingly intricate interdependences [1, 2]. Facing such growing complexities, the traditional *reductionism-based design* approaches, which focus on decomposing and optimizing subsystems, components and their interrelationships within existing products or systems, may face greater difficulties in the search for future innovation.

J. Luo (✉)
Singapore University of Technology and Design, 8 Sompaph Road,
Singapore 487372, Singapore
e-mail: luo@sutd.edu.sg

© Springer International Publishing Switzerland 2016
M.-A. Cardin et al. (eds.), *Complex Systems Design & Management Asia*,
Advances in Intelligent Systems and Computing 426,
DOI 10.1007/978-3-319-29643-2_4

In the meantime, we have also witnessed the invention and surge of many systems of systems [3, 4] in the past three decades, such as the Internet, intelligent transportation systems, smart grids, and the Internet of Things. Such systems of systems normally result from *expansionism-based design* approaches that synthesize existing and separately created systems, instead of decomposing them.

Expansionism-based design may reduce the need to deal with the internal complexity of existing systems, as it focuses more on identifying, integrating, and synthesizing previously existent but unrelated independent systems into a larger holistic system of systems. Therefore, when system designers face super complexities in components and system architectures that constrain their redesign potential via reductionism-based design approaches, expansionism-based design may enable additional out-of-the-box innovation opportunities. Expansionism-based design enables design opportunities for a system of systems that may improve the functionalities and performances of the prior systems, or create novel functions of the new holistic system of systems.

As we anticipate increasing complexities in future products and systems, the importance of expansionism-based design to enable systemic innovation is also expected to grow. This paper aims to formalize and systemize *expansionism-based design* and to provide preliminary guidance for system designers to pursue it. We will review the system theory roots of the reductionism-based versus expansionism-based design approaches, and discuss the advantages and benefits of expansionism-based approaches for the design of complex systems. We will also preliminarily discuss potential concept generation methods to aid in expansionism-based design and methods for the analytics on the collective dynamics and emergent behaviors of resulting systems of systems to support the architecting and management of them.

2 System Theories of Design and Innovation

2.1 *Hierarchies, Decomposition, and Reductionism*

In this paper, we take the classical definition of a "system" as a set of elements in which the behavior of each element has an effect on the behavior of the whole, and the behavior of the elements and their effects on the whole are interdependent [5–7]. The parts of a system will stop functioning and become useless if removed from the system. For example, an eye is a part of the human body but not a system by itself because it will stop functioning after the removal from the body system. Hierarchy is inherent in many types of complex systems [5, 6], such as social, physical, and biological systems. Technological products and systems are also often understood, represented, designed, and managed as a nested hierarchy of recursive subsystems, components, and parts [8, 9].

The design strategies and processes to search for the improvement opportunities of existing products and systems often comply with the hierarchical architectures of existing products and systems. Arthur [10] argued that new technical inventions emerge from an accumulation of previously created components and their functions. Funk [11] showed in detailed cases that many innovations in electronics resulted from the improvements in components and materials. Tushman and Murmann [12] suggested that the evolution trajectory of a technological system is conditioned by the designs and evolution trajectories of its subsystems and components and their coupling at lower levels in the system's nested design hierarchy. Searches for new technologies involve identifying the problem and a recursive process of solving the sub-problems and sub-sub problems. As a result, the hierarchical interdependences between components of a system influence the sequences of the related design processes [13] and the direction of related technological changes [14].

Based on the nested hierarchy view, Henderson and Clark [15] defined architectural innovation as a value-added change in the interrelationships between the system's components, without changing the existing components themselves. This contrasts with component innovations that do not change the relationships between components, but do change the components themselves. Baldwin and Clark [16] focused on system architectures that can enable autonomous and continuous modular innovations within the components. They showed how modular innovations drove technology changes in the personal computer industry. Stone and Wood [17] and Dahmus et al. [18] proposed actionable approaches to decompose a system's main functions into sub-functions and their interdependent relationships, i.e., the functional architecture of a product or system, and identify the modules that can be redesigned without requiring significant changes to other parts of the system.

2.2 Reductionism Versus Expansionism

Given a problem to solve (e.g., improving fuel efficiency and safety) at the level of a complex system (e.g., an automobile), the system designer may implement two alternative design approaches differentiated by the orientations of the search along the system hierarchy: *reduction* versus *expansion*. The system designer may begin by decomposing a system such as an automobile into subsystems, i.e., engine, chassis, body, transmission, suspension, etc., and further into components and parts, none of which will continue to function if removed from the automobile. Then, the designer will seek a more efficient engine, streamline the car body, or change the system architecture (to hybrid-electric, for instance) and re-integrate the new parts with the rest of the system.

Alternatively, the system designer may begin by first conceptualizing a new and larger system (e.g., an intelligent transportation system) that contains the original system of design consideration (i.e., the automobile) and then consider redesigning the automobile in terms of its roles and functions to be played within the transportation system and redesigning the transportation system to improve the utility of the automobile. On that basis, the system designer may design better roads or

infrastructures in the transportation system to allow automobiles to run more efficiently. The designer may also design new policies, such as CAFÉ (Corporate
Average Fuel Economy), which are useful to coordinate the larger system of many
automobiles that are stand-alone complex systems.

The first approach is based on *reductionism*, which decomposes the original
problem into sub-problems to narrow down the focuses of the search for new
solutions. Design is concerned only with the parts within an existing system and the
interdependences among them. In contrast, the latter approach is empowered by
expansionism, which expands the level of thinking and exploration beyond the
original system boundary. It is similarly concerned with interrelationships, but the
relationships are between the original design object, i.e., the focal system, and the
other objects or systems in its environment. The *reductionism-based* and *expansionism-based* design approaches are expected to yield complementary rather than
contradictory results.

There are considerable differences between these two alternative system design
approaches. Reductionism-based design involves *analysis*, whereas the
expansionism-based design involves *synthesis* [7]. Analysis looks into things;
synthesis comes out of things. Analysis focuses on structure, as it reveals how
things work. Synthesis focuses on function, as it reveals why things operate as they
do. Analysis enables us to describe, whereas synthesis enables us to explain.
Analysis yields knowledge, whereas synthesis yields understanding. Analysis is
based upon reductionism; synthesis is based upon expansionism. Increases in
understanding are believed to be obtainable by expanding the systems to be
understood and not by reducing them to their elements.

In the design and innovation literature and practices, *reductionism* has been the
most popular logic for describing technological innovations, as well as the logic
behind many traditional design methods that have been used to pursue technological innovations (see a review in Sect. 2.1). In contrast, *expansionism* has been
largely overlooked, and it can create additional design possibilities for complex
systems. Therefore, when something does not work satisfactorily from within, i.e.,
reductionism, one can think beyond the original system boundary and work out new
solutions from the outside, i.e., expansionism.

3   Expansionism-Based Design

3.1   Definition and Examples

Expansionism-based design encapsulates the focal system by connecting and synthesizing it with other systems, which were previously unrelated, to solve a problem
of the original system, improve it, or generate totally novel and useful functions
from the new and greater system of systems. Figure 1 illustrates such an expansion
of the boundary of thinking and design. The product or service that used to be

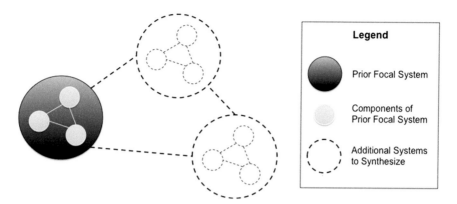

Fig. 1 Expansionism based design to synthesize the prior focal system with other previously unrelated systems

treated and analyzed as an independent system now becomes a "component" of a larger system of systems. Such component systems could maintain a high degree of autonomy while also contributing to other systems. Expansionism-based design extends outward from the existing system design hierarchy so that it becomes a sub-hierarchy of a new and larger hierarchy of the system of systems. It contrasts with reductionism-based design, which focuses on the inward analysis of existing design hierarchy. The expansion makes the analysis of internal complexity less necessary and enables additional out-of-the-box design opportunities.

Some examples of expansionism-based design are presented in Table 1, including a market (the expanded system) compared with an individual person, (the original personal computers (the original system), and *Better Place*'s network of battery swapping stations (the expanded system) for electrical vehicles (the original system), among others. The table lists the original systems that were to be synthesized, their components, the system of systems they form, and the types of linkages that connect them.

Note that both system architecture design [15–18] and expansionism-based design are recursive and are about the design or redesign of "interrelationships". Their differences lie in the differences between *analysis* and *synthesis*. System architecture design arises from redesigning the relationships between non-decomposable or nearly decomposable components that cannot function autonomously or independently if removed from the system. For example, an airfoil does not fly if separated from an airplane. In contrast, expansionism-based design creates new interrelationships for systems to interact with the others that were previously unrelated. A larger system of systems that contains and synthesizes the original systems must be conceptualized, defined, and created. If the new connections among original systems are removed, the original systems could still function independently, just without the value-added from the synthesis of a system of systems.

Table 1 Examples of expansionism-based design

	Original "systems" that later become components of a system of systems	Components of original systems	System of systems	New linkages
1	Human	Productive behaviors	Market	Information
2	Power plants	Boiler, steam turbine	Emission trading system	Information
3	Personal computers	Chips, disk drive, etc.	Internet	Information
4	Personal websites	Advertisements on page	Google AdWords	Information
5	Electrical vehicles	Battery, motor, etc.	Network of battery swapping stations	Materials
6	Thermal power plant, and heat boiler	Boiler, turbine, etc.	Co-generation plant	Materials and energy
7	Cell phone, computer, camera	Chipset, board, keyboard, led, etc.	Smartphone/handset, which synthesizes the functions of phone, computer, internet, camera, etc.	Materials, information, energy
8	Cars and OnStar system	Car components, OnStar equipment, etc.	Vehicle telematics service, including in-vehicle security, turn-by-turn navigation, and remote diagnostics	Information

This difference can be demonstrated using two contrasting examples: the Internet and a telephone network (see Fig. 2). The Internet that connects computers was created via *expansionism-based design* because computers were invented prior to the Internet and had their own independent functionalities. A computer itself was made not of autonomous systems but rather of recursive subsystems and components that provided no value individually. Thus, a computer itself is a system, and the Internet is a system of systems. By contrast, the telephone and telephone network were invented simultaneously as non-separable components of a functional system. A telephone by itself has no value if it is not connected in a network with other telephones. A telephone is a part rather than a system, and the telephone network is a system. Therefore, the invention of the telephone network was not an *expansion* or *synthesis* of telephones. Since its invention, the telephone network as a system (not a system of systems) has experienced many component innovations in the telephones and architecture innovations in the infrastructure that connects the telephones. In brief, the telephone network was invented as a system, and many improvements to the phones were achieved via component innovation, whereas the Internet was invented as a system of systems via expansionism-based design.

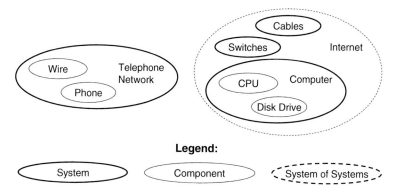

Legend:

System | Component | System of Systems

Fig. 2 System architectures of the telephone network and Internet when they were invented

Another example of such a comparison could be a power plant versus a co-generation plant. A coal power plant solely generating electricity and a boiler solely generating heat are synthesized together in a co-generation plant to simultaneously provide both electricity and heat (from the hot steam exhaust after the turbine process) so that the total efficiency of producing either product is improved through the expansion of design boundaries and synthesis. A co-generation plant is a system of systems, whereas either a coal power plant or a boiler is an independent functional system. The invention of the co-generation plant was expansionism-based. In contrast, reductionism-based design efforts may seek a better chamber structure to improve a boiler or a better steam turbine to improve a power plant.

3.2 Advantages of Expansionism-Based Design

When designing complex products and systems with many interdependent components and intricate interdependences among them, *expansionism-based design* may be more desirable and powerful than *reductionism-based design*. Architectural and modular design efforts search inward through the design hierarchy of a technical system, and therefore, they do not depart from the original concept but rather only reinforce existing functions and mechanisms. Reductionism-based design represents exploitation and is conservative by nature. When the original system or design problem is highly complex and coupled, such approaches may be ineffective in identifying creative solutions because redesigning one component would require the redesign of many other interdependent components due to their intricate coupling.

In such cases, expansionism-based design may be more desirable because it does not require solving the insurmountable challenges that lie within the existing complex system, but aims to discover design opportunities outside the current box of complexities. Expansionism-based design searches outward or upward from the existing design hierarchy of an existing system for design opportunities to link and engage previously unrelated systems. It is therefore more exploratory by nature. To solve rather complex design problems, it will be more advantageous to yield novel out-of-box solutions than reductionism-based design.

For example, the technological evolution of cars has been primarily driven by architectural and modular innovations inside the system of a car, i.e., reductionism-based design. Over time, continual refinements in subsystems, components and their architectures have indeed reinforced the existing dominant design of contemporary cars, and increased the physical and information coupling of the components and parts of a car. The increasing complexity, i.e., interdependences and interlocking relationships between thousands of components and parts within a car, makes it increasingly difficult for either architectural or modular innovations (inside the system) to be achieved. Therefore, expansionism-based design, via searching outside of the existing car design hierarchy and conceptualizing a potential larger system of systems that synthesizes individual cars, may be more likely to create breakthrough designs and innovation opportunities.

For instance, for many decades, electrical vehicles (EVs) have been expected to replace the dominant internal combustion engine (ICE)-powered vehicles. However, the required battery recharging time of 7–8 h and the limited energy and power densities of battery technologies have long hindered the usability and thus mass adoption of EVs. Battery technologies have undergone only limited improvements in the past century and are still insufficient to allow EVs to compete with traditional cars in actual use. Facing the battery bottleneck within an EV, *Better Place*, a venture-backed company founded in 2007, invented a *system of systems* solution that aims to improve the usability of individual EVs without requiring a leap in battery technologies. The solution looks beyond a single car and allows EV drivers to swap exhausted batteries with fully charged ones in 3 min at battery swapping stations, instead of taking 7–8 h to charge the batteries fixed in cars. *Better Place* also planned to integrate the large battery stocks that they hold at swapping stations into intelligent electricity grids; therefore, they can charge the batteries when market electricity price is low and sell excess electricity from its battery stocks to the grid at peak hours. Via optimizing electricity trading, their electricity is expected to be cheaper than that in households, providing an economic incentive for EV owners to drive to *Better Place* battery stations to swap batteries, instead of charging their EVs at home.

Figure 3 illustrates *Better Place*'s design of a system of systems. The electrical vehicles, the battery swapping stations, and the electricity grid are stand-alone systems that can function independently. The synthesis of them was aimed at reconciling the economic and technological constraints on electrical vehicle usability. The battery swapping station in particular connects electrical vehicles and the electricity grid, and this connection benefits both of them. Designers and operators of the electricity grid might view the swapping station as a large electricity storage (sub) system to provide a buffer for the grid, even if not thinking of the electrical vehicles.

Fig. 3 Better Place's expansionism-based design of a system of systems

This particular system-of-systems invention exemplifies *expansionism-based design*, as it expands the design boundary from that of an EV to a larger system of systems that synthesizes EVs, battery swapping stations, battery stocks and electricity grids, to solve the insurmountable economic and technological challenges faced by individual EVs, i.e., the original system. The larger system of systems synthesizes individual electric cars by sharing and swapping their batteries. Therefore, each individual electrical car user can benefit from the shortened charging time from 8 h to several minutes. The prices they pay for electricity would also potentially be lowered because of the financial optimization through the trading of electricity between the electricity grids and the large quantity of batteries held at the network of swapping stations.

This example implies that expansionism-based design can be especially useful and is more likely to yield novel solutions for complex systems such as cars, trains, and airplanes, where modular and architectural innovations have become extremely difficult to achieve due to the prominence of a highly complex and coupled dominant system architecture.

4 Systemic Implications

4.1 Expansive Search Methods

The foregoing section has shown that expansionism-based design offers value for innovation in complex system design. To pursue expansionism-based design, expansive thinking habits and exploratory mindsets of system designers are necessary. More structured methods are also needed to guide systematic exploring and conceiving of new systems of systems, encapsulating the previous system that the designer used to focus on. Several recently developed engineering design methods have the potential to aid in an expansive search for novel concepts of systems of systems. For instance, the infused design [19] framework may empower system engineers to discover new types of relevance between other technologies/systems and the current focal system from a multidisciplinary perspective. The patent search method based on functional analogy [20] may allow system designers to identify and bring new technological systems in other fields of different "distances" from the focal one for potential synthesis together into a new system of systems.

4.2 Coordination, Interfaces and Standards Across Systems

Despite its usefulness, expansionism-based design also faces new and unique challenges in practice. The first is the coordination among prior systems, which were separately designed prior to the synthesis. Therefore, protocols and standards are essential to enable interoperable systems and their effective synthesis [4, 21]. However, setting protocols and standards can be difficult when the designers, operators, and other stakeholders of the component systems have heterogeneous interests and incentives [3, 4]. In the example of *Better Place*, to achieve sufficient economy of scale, the batteries stored and swapped at *Better Place*'s stations need to share technical standards with those of the electrical vehicles designed and made by different automobile manufacturers, e.g., Tesla, Toyota, and General Motors.

However, the EV manufacturers would be naturally concerned that *Better Place* will dominate the system of systems, marginalize them, and capture most of the value created, if it sets the technical standards. In fact, *Better Place* successfully convinced and collaborated with Nissan to share battery standards. *Better Place* went bankrupt in 2013 due to financial and operational challenges. Interestingly, around the same time of the bankruptcy of *Better Place*, Tesla Motors started to offer a 90-s battery-swapping option, similar to that of *Better Place*, in its super-charging network. Compared with *Better Place*, Tesla Motors has advantages in realizing the system of systems because it designs and produces its own electrical vehicles, batteries, and battery swapping systems in house. Coordination and standard setting is relatively easier.

This case illuminates alternative approaches to address the challenges to setting and implementing interface standards across agents. For instance, one approach is to form alliances, such as the one between *Better Place* and Nissan, and industry consortiums or associations, such as the *Internet Engineering Task Force* that develops and manages technical standards of Internet technologies, to facilitate coordination and collaboration. Alternatively, the designer that used to focus on a component system may further expand its design, operational and business scopes to cover other component systems as well as the interfaces among them, such as Tesla expanding its scope from EVs to the supporting infrastructures.

4.3 Prediction of Emerging Behaviors

Because the component systems of a system of systems are autonomous by nature, to some degree, after the synthesis, there might be *collective* and *emergent* behaviors from the interactions of such semi-autonomous systems and their designers and operators [3, 4]. One example is the rapid diffusion of a computer virus across Internet. Therefore, the success of expansionism-based design also requires the development of the capabilities to predict, assess, and manage potential emergent behaviors of the resulting system of systems. Recent studies have suggested agent-based models to represent a system of systems and analyze the interaction dynamics among agents [22]. Graph theory and complex network analysis may also be used to predict the propagation of information or design changes across autonomous systems and aid in the design of the architecture of the new system of systems [23]. More research and development is still needed to allow such general models and analyses to generate practical insights on the design and management of systems of systems.

4.4 Impact on Innovation and Competition Landscapes

Expansionism-based design may also affect innovation and competition dynamics in related markets and industries, in both undesirable and desirable ways. For instance, successful expansionism-based design stimulates modular and architectural innovations inside the original systems, while the synthesis demands changes in the functional goals, design mechanisms, and parameters of these now component systems. For example, the diffusion of smart phones and mobile internet has resulted in demand for smaller and more powerful processors, smaller and faster communication chips, more efficient memories, larger handset screens, and more advanced operating systems, which are most likely to be improved via component and architectural innovations. Clearly, engineers and firms specialized in such component niches will also benefit from the pull from the emergence of a new system of systems.

New systems of systems may also increase the market demands for the original systems that are now synthesized. For example, if the battery-swapping infrastructure prevails, the use feasibility of electric cars will be improved, and the demand for electric cars will be stimulated. However, in other cases, the growth of the new system of systems may limit or replace the needs for the prior systems that it now encapsulates, thus threating the designers and firms specialized in them. For example, smart phones have increasingly encroached upon the market shares of basic mobile phones as well as digital cameras. Consumers are less interested in buying standalone basic cell phones and cameras, as they can use the corresponding functions that have been synthesized into their smart phones. The firms that were specialized in designing such prior systems need to adapt and make their system designs compatible with the system of systems, and be able to harvest the value created from the synthesis with others.

Therefore, when pursuing expansionism-based design, it is necessary and important for the designers and companies to evaluate the potential social-technological and competitive impacts of the anticipated system of systems and to prepare themselves with relevant capabilities potentially required to address those changes. Many of these issues transcend the basic assumptions in traditional engineering design and require new ways of thinking and new system design and management methods, drawn more upon system sciences, complexity studies, and evolutionary theories. Well-prepared system designers and firms may better capture the value created by expansionism-based design and the resulting system of systems and minimize the associated risks of undesirable resultant changes in technologies and markets.

5   Closing Remarks

The purpose of this paper is to advocate expansionism-based design, which has been largely ignored or not been explicitly used in design and innovation literature and practices, compared with reductionism-based design. Expansionism-based design complements reductionism-based design and can enable additional innovation opportunities by synthesizing existing systems and creating systems of systems. We argue that it is most advantageous and powerful for designing and managing highly complex systems, for which opportunities for component and architecture innovations are limited by the intricate interdependence architecture and component performances. Therefore, as we expect the complexity in contemporary technologies and systems to monotonically rise in the future [1], the importance of expansionism-based design for system innovation is also expected to increase.

To aid in expansionism-based design, some recently developed engineering concept generation methods can be leveraged, for example, infused design and design-by-analogy patent search methodology. In the meantime, expansionism-based design and its resulting systems of systems may require cross-agent coordination and setting standards and protocols for the interfaces of the component systems. To reconcile potential conflicts of interests and synchronize incentives of agents designing different component systems, alliances and consortiums may be beneficial for the participants. Alternatively, to avoid the agent problem, the designer of certain component systems may expand its scope to integrate the design of other component systems and the interfaces among them. In addition, the resulting systems of systems may have unpredictable, either desirable or undesirable, impacts to the existing technological and competition landscapes. Agent-based modeling and complex network analysis are potentially useful for system designers to predict the collective dynamics of the potential system of systems for evaluation and aid in system architecting.

In general, to harvest the benefits from expansionism-based design and the resulting system of systems, system designers and their companies need to develop relevant capabilities and approaches for the boundary-expanding search for new system of systems concepts, for predicting and managing emergent behaviors and collective dynamics of the system of systems, and for architecting the system of systems. We hope this paper can be seen as an invitation for more academic research into the development of theories and methods to aid in expansionism-based design and the analysis, architecting, and management of the resulting system of systems.

References

1. de Weck, O., Roos, D., Magee, C.L.: Engineering Systems: Meeting Human Needs in a Complex Technological World. The MIT Press, Cambridge (2011)
2. Ameri, F., Summers, J., Mocko, G., Porter, M.: Engineering design complexity: an experimental study of methods and measures. Res. Eng. Design **19**, 161–179 (2008)
3. Nicola, R.N., Fitzgerald, M.E., Ross, A.M., Rhodes, D.H.: Architecting systems of systems with ilities: an overview of the SAI method. Procedia Comput. Sci. **28**, 322–331 (2014)
4. Butterfield, M.L., Pearlman, J.S., Vickroy, S.C.: A system-of-systems engineering GEOSS: architectural approach. IEEE Syst. J. **2**, 321–332 (2008)
5. Alexander, C.: Notes of the Synthesis of Form. Harvard University Press, Cambridge (1964)
6. Simon, H.A.: The Sciences of the Artificial, 3rd edn. The MIT Press, Cambridge (1996)
7. Ackoff, R.L.: Creating the Corporate Future. Wiley, New York (1981)
8. Clark, K.B.: The interaction of design hierarchies and market concepts in technological evolution. Res. Policy **14**, 235–251 (1985)
9. Christensen, C., Rosenbloom, R.S.: Explaining the attacker's advantage: Technological paradigms, organizational dynamics, and the value network. Res. Policy **24**, 233–257 (1995)
10. Arthur, B.W.J: The structure of invention. Res. Policy **36**, 274–287 (2007)
11. Funk, J.: Systems, components, and technological discontinuities: the case of magnetic recording and playback equipment. Res. Policy **38**, 1079–1216 (2009)
12. Tushman, M.L., Murmann, J.P.: Dominant designs, technology cycles and organizational outcomes. Res. Organiz. Behav. **20**, 231–266 (1998)
13. Marple, D.L.: The decisions of engineering design. IEEE Trans. Eng. Manag. **2**, 55–71 (1961)
14. Rosenberg, N.: Perspectives on Technology. Cambridge University Press, Cambridge (1961)
15. Henderson, R.M., Clark, K.B.: Architectural innovation: The reconfiguration of existing product technologies and failure of established firms. Adm. Sci. Q. **35**, 9–30 (1990)
16. Baldwin, C.Y., Clark, K.B.: Design Rules, Volume 1: The Power of Modularity. The MIT Press (2000)
17. Stone, R.B., Wood, K.L., Crawford, R.H.: A heuristic method to identify modules from a functional description of a product. Des. Stud. **21**, 5–31 (2000)
18. Dahmus, J.B., Gonzalez-Zugasti, J.P., Otto, K.N.: Modular product architecture. Des. Studies **22**, 409–424 (2001)
19. Shai, O., Reich, Y.: Infused design: 1 Theory. Res. Eng. Design **15**, 93–107 (2004)
20. Fu, K., Murphy, J., Fu, K., Yang, M., Otto, K., Jensen, D., Wood, K.: Design-by-analogy: experimental evaluation of a functional analogy search methodology for concept generation improvement. Res. Eng. Design **26**, 77–95 (2015)
21. Corsello, M.A.: System-of-systems architectural considerations for complex environments and evolving requirements. IEEE Syst. J. **2**, 312–320 (2008)
22. Acheson, P., Dagli, C., Kilicay-Ergin, N.: Model based systems engineering for system of systems using agent-based modeling. In: Conference on Systems Engineering Research, Atlanta, GA, 19–22 March (2013)
23. Newman, M.E.J., Barabási, A.L., Watts, D.J.: The Structure and Dynamics of Networks. Princeton University Press, Princeton (2006)

A General Framework for Critical Infrastructure Interdependencies Modeling Using Economic Input-Output Model and Network Analysis

Jiwei Lin, Kang Tai, Robert L.K. Tiong and Mong Soon Sim

Abstract Critical infrastructures are essential systems for the continuous functioning of modern society. These infrastructures are closely linked and dependent on one another, and these interdependencies need to be modeled in order to analyze the disruptions and vulnerabilities of critical infrastructure networks as a whole. In this study, a Leontief input-output (I/O) model and a critical infrastructure network system model will be developed. The I/O model serves to describe the interdependencies among the critical infrastructure systems while a critical infrastructure network will describe the network structure of a real physical infrastructure network of concern. An infrastructure network of Singapore is constructed to exhibit the use of network analysis with the I/O model. The study aims to create a general framework for economic impact studies that can be used in any country with a national I/O table and the network structure of a critical infrastructure of concern. This framework can ultimately provide insights to the fragility of the economic system to various forms of disruption and provide guidance to policymakers.

J. Lin (✉)
Institute of Catastrophe Risk Management (ICRM), Interdisciplinary Graduate School, Nanyang Technological University, 50 Nanyang Ave, Singapore 639798, Singapore
e-mail: linj0068@ntu.edu.sg

K. Tai
School of Mechanical and Aerospace Engineering, Nanyang Technological University, 50 Nanyang Ave, Singapore 639798, Singapore
e-mail: mktai@ntu.edu.sg

R.L.K. Tiong
School of Civil and Environmental Engineering, Nanyang Technological University, 50 Nanyang Ave, Singapore 639798, Singapore
e-mail: clktiong@ntu.edu.sg

M.S. Sim
DSO National Laboratories, 20 Science Park Drive, Singapore 118230, Singapore
e-mail: smongsoo@dso.org.sg

© Springer International Publishing Switzerland 2016
M.-A. Cardin et al. (eds.), *Complex Systems Design & Management Asia*,
Advances in Intelligent Systems and Computing 426,
DOI 10.1007/978-3-319-29643-2_5

59

1 Introduction

The world's social and economic stability is largely dependent on a reliable and continuous flow of goods and services provided by infrastructure such as the electricity and the Internet. It would be more unimaginable nowadays if a country were to live in an environment without electricity than it would have been about 50 years ago. The heavy reliance on goods and services forces some specific infrastructure to be fully operational at a 100 % uptime.

In the modern world, critical infrastructure is a network of independent, man-made systems and processes that function collaboratively and synergistically to produce and distribute a continuous flow of goods and services [1] essential to the defense, economic, security and the smooth functioning of the government and society as a whole. Critical infrastructure is a term usually used to describe assets that are essential for the functioning of a society and economy.

Critical infrastructure has been in the spotlight in recent years due to the immense amount of attention captured through different events ranging from man-made disasters (terrorist attack on 11th September 2001) to natural disasters (2005 Hurricane Katrina, 2011 Tohoku Earthquake) that caused damage to the well-being of society. These events usually result in a domino effect propagated to other infrastructures. A failure in one of the critical infrastructure will have enormous impact on one or more of the following: health, safety, security, economic, social well-being and the functioning of the government. It has become unacceptable for critical infrastructure to fail as there will be severe inconvenience to society and financial losses will be huge.

A common observation from the disasters are that they are able to affect different infrastructures. In fact, they are the consequences of the interaction between different infrastructures, due to their interconnected nature. It is logical to think that real networks do not work independently. They require interaction with some other infrastructure network for itself to be operational.

Many countries have also turned their attention to the consequences of disaster. The United States is the first country to make a stand in focusing on critical infrastructure protection with the publication of "Critical Foundations: Protecting America's infrastructure" [1] in 1997. The finding of this report emphasize the possibility of devastating effect that are caused by little knowledge of critical infrastructure. The European Council started on a similar focus in 2004, which is published in the report called "Critical Infrastructure Protection in the fight against terrorism" [2] which lead to further extension of assessment required. However, they have noticed that characterizing the interdependencies among infrastructures is still a major challenge in the recent study as such interdependencies may be characterized and measured from different perspectives and views from different stakeholders and researchers.

In this study, we will be using input-output (I/O) to model interdependencies. It has been used to model the interactions among sectors in the economy [3, 4] and forecast the impacts of the changes in one part of the economy on the others.

In this paper, we propose a method to explore the interdependencies of critical infrastructures using a modified I/O model and incorporating a critical infrastructure network which describes a real physical infrastructure network. The I/O model serves to describe the interdependencies among the critical infrastructure systems while a critical infrastructure network will describe the network structure of a real physical infrastructure network of concern. To analyze the magnitude and extent of the interdependencies of critical infrastructures, an interdependency matrix based on the Singapore National I/O Table is formulated. The telecommunication infrastructure network of Singapore is constructed to exhibit the use of network analysis in conjunction with the critical infrastructure interdependencies model developed. Due to the ease of use of publicly available data, this framework can potentially serve as an interdependency model template for researchers and critical infrastructure stakeholders that are looking at critical infrastructure interdependency.

2 Infrastructure Interdependencies and Critical Infrastructure Network

2.1 Input-Output Model as Critical Infrastructure Interdependencies

The I/O model is a quantitative economic technique that represents the interdependencies between different branches of a national economy or different regional economies [5]. As the I/O model is fundamentally linear in nature, it lends itself to rapid computation as well as flexibility in computing the effects of changes in demand. Changes in demand act as a "shock" to the economy and the I/O analysis tells us the full impact of that shock on both the sector in question and the other sectors to which it is linked.

Leontief's I/O model is constructed from an observed set of data (e.g. National I/O table) for one economic area, for example, a country's economy. Economic activities in the country can be categorized into a number of sectors (e.g. utilities, transportation, etc.). The necessary data are the flows of products from each of the sectors (as seller) to each of the sectors (as buyer). This type of inter-industry (or inter-sectoral) transactions are measured for a period of time period (usually one year) and in monetary terms.

The nature of I/O model provides a possible way to analyze the economy as an interconnected system of industries that can be directly or indirectly affected by one industry sector or another. I/O models trace the linkages from the raw material stage to the sale of the product as goods. This advantage enables the estimation of economic impacts of any changes to the economy based on these inter-industry transactions. The ability to analyze and capture the economy's direct and indirect reaction to change in the economic environment (any perturbation) makes I/O unique. This advantage is helpful in the current research as critical infrastructure network analysis requires a model that is able to link infrastructures together, according to interdependencies [6].

The assumption being made in this proposed method is that monetary transactions from the I/O tables reflect the physical and logical (in terms of economic) interdependencies. The greater the monetary transactions between two industry sectors in an economy, the more interconnected they are.

The mathematical formulation is based on demand-pull I/O quantity model [7]. If x_i represents the total output (production) of sector i, z_{ij} represents the monetary values of the transaction between pairs of sector (from sector i to sector j) and final demand f_i represents total final demand for sector i's product. We may write a simple equation accounting for the way in which sector i sells its product as a form of input to other sectors and to the final demand:

$$x_i = z_{i1} + z_{i2} + z_{i3} + \cdots + z_{in} + f_i = \sum_{j=1}^{n} z_{ij} + f_i \tag{1}$$

In the I/O model, the technical coefficient a_{ij} is defined as the value of product i required as input to produce a unit value of product j, and can be expressed as:

$$a_{ij} = \frac{z_{ij}}{x_j} \tag{2}$$

Substituting Eq. (2) into (1), the sales of the output of each of the n sectors can be written in matrix form as in Eq. (3), simplified to Eq. (4) and rearranged to Eq. (5):

$$\begin{pmatrix} x_1 \\ x_2 \\ \vdots \\ x_n \end{pmatrix} = \begin{pmatrix} a_{11} & a_{12} & \cdots & a_{1n} \\ a_{21} & a_{22} & \cdots & a_{2n} \\ \vdots & \vdots & \ddots & \vdots \\ a_{n1} & a_{n2} & \cdots & a_{nn} \end{pmatrix} \begin{pmatrix} x_1 \\ x_2 \\ \vdots \\ x_n \end{pmatrix} + \begin{pmatrix} f_1 \\ f_2 \\ \vdots \\ f_n \end{pmatrix} \tag{3}$$

$$X = AX + F \tag{4}$$

$$(I - A)X = F \tag{5}$$

The matrix A is usually called the direct requirement matrix, which represents the technological state of the economy for producing products for all sectors.

And finally, applying standard matrix algebra to solve for X (the output for all sector) in Eq. (5):

$$X = (I - A)^{-1}F \tag{6}$$

Equation (6) represents the solution of the I/O model. X represents the output of all sectors due to the effect of the final demand F. $(I - A)^{-1}$ is known as the

Leontief inverse or total requirement matrix. This inverse will also be referred to as **interdependency matrix** that controls the direct and indirect effects due to any change in final demand F. The interdependency matrix will be illustrated as I/O interdependency model as shown in Fig. 1. Since Eq. (6) is a linear equation, it can also be represented by:

$$\Delta X = (I - A)^{-1} \Delta F \qquad (7)$$

where the change in output X is directly proportional to the change in final demand F. ΔF is very important for our study as I/O models are based on the assumption that export demand (or the ability of industries to sell to the external economy) is the engine that generates activity in the regional economy. Changes in final demand, ΔF, infuse local industries with new funds (in the case of positive change in demand), which increase output X [8]. However, a negative change in final demand F will vice versa lead to decrease in output X. This is the main attribute of I/O model where a perturbation, ΔF, to the critical infrastructure will result in damages in the form of decreased output X to all critical infrastructure connected within the I/O model. Due to the equilibrium assumption of the Leontief model, the economic losses are typically estimated on an annual basis. Hence, for smaller time resolutions, it is assumed that the losses are evenly distributed throughout the year [9]. Another assumption that arises from considering a relatively short period of time is that all the considered industries in the analysis reach their steady-state values within the considered period of time. However, in practice, it may take longer for impact to cascade into other critical infrastructures.

There are advantages and flaws in using I/O model in critical infrastructure modelling. The advantages are the accessible and suitable data available publicly, linearity of I/O model which lends itself to rapid computation as well as flexibility in computing the effects of changes in demand, and good representation of the inter-industry transaction which in turn represents the interdependencies among all industry sectors in a nation [5]. However, a clear flaw identified is that the I/O model is unable to describe real physical infrastructure networks. For example, the magnitude of I/O value does not always reflect the criticality of the resource link. A lower magnitude does not imply that the receiving sector depend less on the supplying sector for operation. An example is utilities used by a sector, which usually is not a major operating cost, but yet vital for its operation. Furthermore, firms may have mitigation measures if a certain resource is known to be critical for its operation. And the adequacy of any mitigation measures is not captured via the I/O value. Therefore, an effort to link real physical infrastructure network and I/O interdependency model (which is generated from interdependency matrix) is being attempted in this paper.

Based on the Singapore economic I/O table (Singapore National I/O Table 2007), the aggregation of the 136 sectors into respective critical infrastructure sectors (9 main critical infrastructure sectors + 1 sector that is an aggregation of the remaining ungrouped industry sectors) was done (Appendix A). The 136 sectors have been aggregated into 10 sectors based on reference [10] and with references to

[11] due to their nature of importance exhibited in Singapore context. The new aggregated I/O model is then used to generate the I/O interdependency matrix (Appendix B) that depicts the links between critical infrastructure sectors as shown in Fig. 1.

Through the interfacing of real physical infrastructure network and I/O interdependency model, analyzing one critical infrastructure in I/O interdependency model becomes possible. The advantage would be that the use of real physical infrastructure network would be more potent in modelling real world events which lead to cascading damages to other sectors that could not be established in real physical network but could be represented as sectors among the I/O interdependency model. Thus, with the interfacing, the effect and extent of impact to other sectors can be established if an event that causes damage to real physical network occurs.

The I/O interdependency model serves the purpose of describing physical linkage of the 10 critical infrastructure sectors which is also a good way of representing their connection. A flaw of I/O interdependency model is, the details of each sector is still ultimately not established if there is no supporting data available. It will only be able to provide a macro view of what will happen to other sectors as a whole if a known physical infrastructure has been affected.

2.2 Critical Infrastructure Network

In this subsection, a critical infrastructure network is chosen for network analysis. The network corresponds to the main linkages of the telecommunication network in Singapore. The network comprises of two sets of elements: central offices and transmission links. Figure 1 displays the topology of the sample telecommunication network developed in QGIS [12]. In this study, we will consider a disruption to a node as node failure to simplify the study. A node failure causes all links that are connected to that node to be disabled.

2.3 Fusion of Interdependencies and Critical Infrastructure Network

Upon any failure to the physical infrastructure network, the network performance for the physical infrastructure will be measured. Two network performance metrics will be used in this study. The two metrics are (1) efficiency of network [13, 14] and (2) network average clustering coefficient [15–17]. There have not been any studies to determine which of the two metrics is more appropriate as a measure of the network performance, hence both metrics are used in this work.

The efficiency of network is given by

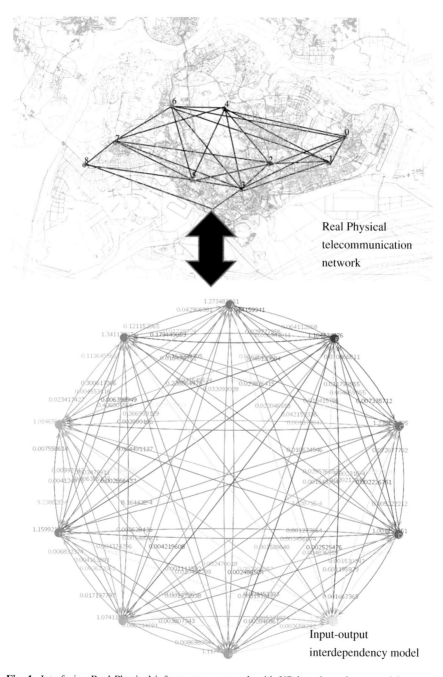

Fig. 1 Interfacing Real Physical infrastructure network with I/O interdependency model

$$\eta = \frac{1}{n(n-1)} \sum_{i,j \in V} \frac{1}{d_{ij}} \tag{8}$$

where n represents the number of nodes in a network, and d_{ij} represents the shortest path length between nodes i and j.

Average clustering coefficient of a network [18] is defined as the average of the clustering coefficient of all individual nodes in the network, which is given by

$$\eta = \frac{1}{n} \sum_{i=1}^{n} \frac{\text{number of triangles connected to node } i}{\text{number of triplets centered on node } i} \tag{9}$$

where n represents the number of nodes in a network.

The disruption (e.g. terrorist attack) will be simulated and the network performance of the physical infrastructure network will be measured. The losses of network performance upon node failure in physical infrastructure network can be calculated as

$$e_{loss} = \left(1 - \frac{\eta_{new}}{\eta_{initial}}\right) \times 100\% \tag{10}$$

where $\eta_{initial}$ is the initial network performance and η_{new} is the new network performance. Both network performance must be calculated according to the selected network performance metric. If the performance of the network after a node failure is lower, the removed node is critical (e_{loss} will be positive).

When the failure of telecommunication network happens, it will result in failure to other central office of the same telecommunication network. The failure of physical infrastructure are being captured within a fixed time frame (e.g. 1 day) and the losses of the particular infrastructure due to node failure can be captured in the form of contribution to the economy as a whole in a day. We assume that for each day of failure of a physical infrastructure, it will incur losses in the form of:

$$w_{loss} = \frac{x_i}{365} e_{loss} \tag{11}$$

where w_{loss} represents the monetary loss for an infrastructure sector, x_i represents the output of the sector i in a year, e_{loss} represents the losses of network performance upon failure in the physical infrastructure. The w_{loss} will be used as ΔF in the I/O model. The losses of network performance will be able to propagate across to the other industry sectors and the total damage to all industry sectors can be evaluated. A complete exhaustive search for all cases of node failure can be performed to find the worst case scenario for node failure for the selected number of nodes.

2.4 Strength and Limitations

The framework proposed utilize I/O table which is a reliable and open-source data source. Interesting observations can emerge from the interdependencies among different sectors. However, there are some limitations with this framework. There are no cross references to some particular critical infrastructure as I/O model is an economic base model and some critical infrastructures do not make up a large proportion of monetary output in the national economy. An example for this scenario is electricity which holds smaller economic value as compared to other sectors but is very critical to a nation. Some critical infrastructure like military defense which is traditionally included as critical infrastructure is not included as a sector in I/O table as it does not technically contribute to the national economy but is critical to the defense of a nation. Finally, network performance metric are the next best approach, under the assumption that users do not have much knowledge about the network such as capacity, routing protocol, hierarchy and other data.

3 Network Simulation and Discussion

In this section, we will utilize computer simulation to analyze the network response of the critical infrastructure network to disruption due to, for example, a targeted terrorist attack. In the simulation, the critical infrastructure network is first constructed in QGIS and imported into Netlogo, which is a multi-agent programming language and integrated modeling environment [19]. Netlogo is employed due to its ease of use and its powerful simulation feature which simulates 'agents' freely. Figure 2 shows the graphical representation of the Netlogo programme interface.

In the Netlogo programme, any failure in the critical infrastructure network will disconnect all direct links to the failed node. This failure scenario is implemented to simulate the different combination of node failures and their impact on all critical infrastructure sectors linked in I/O interdependency model [20–22]. The simulation contains One-node failure, Two-node failure and Three-node failure. These failure scenarios attempt to simulate node failures in a particular time frame, for this case, the time frame is a one full day scenario. Two network performance metrics, (1) Efficiency of network and (2) Network average clustering coefficient, will be used to assess the network performance, η, of the critical infrastructure network. A complete exhaustive search for all cases of node failure will be performed to find the worst case scenario upon the failure of one, two and three nodes. e_{loss} will be calculated based on the two different network performance metrics and thereafter, w_{loss} can be calculated, and will be used as ΔF for I/O model to assess the impact to all sectors, ΔX, due to the node failures.

Tables 1, 2 and 3. show the results of the top 3 worst case scenarios for one-node, two-node and three-node, failures respectively. Different network performance metrics will produce different top 3 worst case scenarios due to the

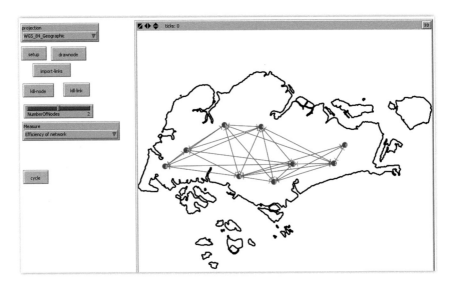

Fig. 2 Netlogo programme interface for modelling and simulation of node failure of critical infrastructure network with 9 nodes

Table 1 Results of the top 3 worst case scenarios for one-node failure

Network performance scheme	Failure scenario	$\eta_{initial}$	η_{new}	e_{loss}	Overall impact, ΔX ($ Millions)
Efficiency of network	[0]	1.28	1.18	7.76	6.94
	[7]	1.28	1.25	2.17	1.94
	[8]	1.28	1.25	2.17	1.94
Network average clustering coefficient	[5]	0.793	0.717	9.64	8.62
	[4]	0.793	0.730	7.99	7.14
	[7]	0.793	0.796	−0.342	−0.306

Table 2 Results of the top 3 worst case scenarios for two-node failure

Network performance scheme	Failure scenario	$\eta_{initial}$	η_{new}	e_{loss}	Overall impact, ΔX ($ Millions)
Efficiency of network	[0 1]	1.28	1.14	10.6	9.44
	[0 3]	1.28	1.14	10.6	9.44
	[0 7]	1.28	1.14	10.6	9.44
Network average clustering coefficient	[4 5]	0.793	0.571	28.0	25.0
	[1 5]	0.793	0.576	27.4	24.4
	[3 5]	0.793	0.576	27.4	24.4

Table 3 Results of the top 3 worst case scenarios for three-node failure

Network performance scheme	Failure scenario	$\eta_{initial}$	η_{new}	e_{loss}	Overall impact, ΔX ($ Millions)
Efficiency of network	[0 1 3]	1.28	1.07	16.5	14.8
	[0 7 8]	1.28	1.07	16.5	14.8
	[1 3 5]	1.28	1.10	13.9	12.4
Network average clustering coefficient	[1 4 5]	0.793	0.333	58.0	51.8
	[3 4 5]	0.793	0.333	58.0	51.8
	[4 5 8]	0.793	0.444	44.0	39.3

different nature of the network performances metric being used. Further studies will be required to

validate the suitability of the metric. For efficiency of network performance metric, node 0 is found to be the most critical node in one-node failure while node 5 is found as the most critical node in one-node failure for network average clustering coefficient. A note on the results of the top 3 worst case scenarios for one-node failure is the failure of node 7 for network average clustering coefficient. The resultant impact is a negative value which can be explained as failure of the node actually increase the network performance of the critical infrastructure. This is not illogical from network analysis point of view as some nodes are not as significant (within the particular network performance metric) as other nodes and therefore removing them actually helps in improving the performance of the network. In reality, this is not always the case as the failure of that particular node will still result in significant impact on the critical infrastructure and on the whole of the nation. Therefore in our analysis, we will focus on the positive overall impact and the most positive overall impact failure scenario will be classified as the most significant.

From the two-node failure results in Table 2, the node pairs in the top 3 scenarios for efficiency of network scheme contain node 0 which is also the most critical node for one-node failure in Table 1. In network average clustering coefficient, the node pairs in the top 3 scenarios contain node 5 which is also the most critical node for one-node failure under the same scheme in Table 1. It may seem logical that the top overall impact of two-node failure will always contain the top few critical node in one-node failure, there is a need for further studies on the node-pair combination that will cause the greatest overall impact to a nation. It may be a network attribute dependent on the complexity of the critical infrastructure network and therefore network simulation is recommended before any conclusion can be made.

Results from Table 3 also display similar findings in terms of criticality of node. For efficiency of network under three-node failure, nodes [0 1 3] and nodes [0 7 8] will cause to the greatest overall impact while for network average clustering coefficient, nodes [1 4 5] and nodes [3 4 5] will cause the greatest overall impact. With the capability of the simulation, this framework will be significantly useful for researchers and stakeholders.

4 Summary and Conclusion

Critical infrastructures are different in form and are interconnected. This brings about the complexity of modeling them all together. By using modelling and simulation, these interdependencies can be captured and analyzed systematically. In the area of research of critical infrastructure protection, the information about interdependencies is normally classified with high confidentiality and not available in the public domain. This work explores the possibility of using publicly available data for critical infrastructure modelling and research. The aim of this framework is to provide insights to the fragility of the economic system to various forms of disruption and provide guidance to policymakers. The I/O interdependency model represents the cascading effect of some failure throughout the economy through the interdependency modelled. This framework serves as a prediction tool for pre-disaster deployment, reference for decision support and provide forecast based on a model. The current framework is a generic one and can be used for all countries with publicly available I/O table. The current network performance metric used may not be the best performance metric to use and therefore is an interesting area to look into. The next stage of the research will be on the construction of two or more known critical infrastructure networks and the simulation of disruption by similar methods. The overall damage propagated through the I/O interdependencies will create unique insight of overall economic damage done to the whole of a nation's critical infrastructures.

Appendix A: Aggregation of Economic Sector into Respective CI

No	Main CI sector	Relevant sector in I/O table by ID	Name of the sector
1	Banking and finance	101	Life insurance
		102	General and other insurance
		103	Banks and finance companies
		104	Other financial services
		105	Fund management
2	IT	100	Communications
		109	Information technology
3	Energy	77	Electricity
4	Water	79	Water
5	Transportation	88	Passenger transport by land
		95	Freight transport by land
6	Health	126	Medical and health services
		127	Environmental health services

(continued)

(continued)

No	Main CI sector	Relevant sector in I/O table by ID	Name of the sector
7	Food supply	1	Nursery products
		2	Other agriculture
		3	Livestock
		6	Food preparations
		7	Bread, biscuits and confectionery
		8	Sugar, chocolate and related products
		9	Oils and fats
		10	Dairy products
		11	Coffee and tea
		12	Other food products
		13	Soft drinks
		14	Alcoholic drinks and tobacco products
		29	Food chemicals and additives
		86	Food and beverage services
8	Aviation security	92	Air transport
		93	Supporting services to air transport
		94	Airport operation services
9	Maritime security	89	Water transport
		90	Supporting services to water transport

Appendix B: Critical Infrastructure Interdependency Matrix

Sales by industry\sales by commodity	Banking and finance	IT	Electricity	Water	Transport	Health	Food supply	Aviation security	Maritime security	Others
Banking and finance	1.273483	0.01416	0.029777	0.045121	0.05253	0.032093	0.03165	0.0185	0.017182	0.042907
IT	0.064113	1.104235	0.010867	0.011791	0.035275	0.042159	0.022047	0.029876	0.010206	0.026942
Electricity	0.004869	0.007336	1.238137	0.072078	0.016593	0.020213	0.029577	0.010625	0.001877	0.015416
Water	0.00058	0.002227	0.005222	1.001157	0.001196	0.00153	0.004837	0.001274	0.000364	0.001545
Transport	0.003497	0.002525	0.001196	0.001467	1.184705	0.002658	0.005372	0.003193	0.002001	0.003589
Health	0.00247	0.002487	0.004453	0.002847	0.005674	1.11771	0.008699	0.003807	0.001364	0.010483
Food supply	0.005481	0.00422	0.001112	0.001923	0.003267	0.005235	1.074119	0.017198	0.003622	0.004374
Aviation security	0.003473	0.002866	0.000816	0.002628	0.002634	0.004163	0.006832	1.159922	0.000924	0.004125
Maritime security	0.004553	0.006399	0.002999	0.004491	0.032544	0.006386	0.009927	0.007559	1.084657	0.023417
Others	0.121153	0.179449	0.163099	0.238865	0.362797	0.266933	0.406806	0.300617	0.113646	1.341172

References

1. The President's Commission on Critical Infrastructure Protection.: Critical Foundation: Protecting America's Infrastructure. USA (1997)
2. European Commission.: Critical Infrastructure Protection in the Fight Against Terrorism. Belgium (2004)
3. Leontief, W.: Input-Output Economics. 2nd edn. Oxford University Press, New York (1986)
4. Shaik, A., Tonak, E.A.: Measuring the Wealth of Nations: The Political Economy of National Accounts. Columbia University Press, New York (1994)
5. Baumol, W.J., ten Raa, T.: Wassily Leontief: in appreciation. Eur. J. Hist. Econ. Thought, Taylor & Francis Journals **16**(3), 511–522 (2009)
6. Rinaldi, S.M., Peerenboom P.J.: Identifying Understanding and analyzing critical infrastructure interdependencies. IEEE Control Syst. Mag. 11–25 (2001)
7. Miller, R.E., Blair, P.D.: Input-Output Analysis: Foundations and Extensions. Cambridge University Press, New York (2009)
8. Liu, Z., Ribeiro, R., Warner, M.: Comparing child care multipliers in the regional economy: analysis from 50 states. Cornell University, Department of City and Regional Planning, Ithaca, NY (2004)
9. Anderson, C.W., Santos, J.R., Haimes, Y.Y.: A risk-based input-output methodology for measuring the effects of the august 2003 northeast blackout. Econ. Model. Disaster Impact Anal. **19**(2), 183–204 (2007)
10. Cf. National Security Coordination Centre.: The Fight Against Terror—Singapore's National Security Strategy (2004). www.nscs.gov.sg/public/download.ashx?id=48. Accessed 16 July 2015
11. Speech by Senior Minister of State For Law and Home Affairs Ho Peng Kee at the Monoc Seminar, 22 March 2002. http://mha3.a8hosting.com/news_details.aspx?nid=ODc2-elEze6y%2bGUo%3d. Accessed 16 July 2015
12. QGIS Development Team.: QGIS Geographic Information System. Open Source Geospatial Foundation (2009) http://qgis.osgeo.org
13. Crucitti, P., Latora, V., Marchiori, M.: Model for cascading failures in complex networks. Phys. Rev. E **69**, 045104 (2004)
14. Wang, S., Hong, L., Ouyang, M., Zhang, J., Chen, X.: Vulnerability analysis of interdependent infrastructure systems under edge attack strategies. Saf. Sci. **51**, 328–337 (2013)
15. Mao, et al. Z.: Interdependency Analysis of Infrastructures. In: Proceedings of the 6th International Symposium on Neural Networks: Advances in Neural Networks—Part III (ISNN 2009) (2009)
16. Brust, M., Rothkugel, S.: Small-worlds: Strong clustering in wireless networks. In: First International Workshop on Localized Algorithms and Protocols for Wireless Sensor Networks, LOCALGOS, USA (2007)
17. Holmgren, Å.J.: Using graph models to analyze the vulnerability of electric power networks. Risk Anal. **26**(4), 955–969 (2006)
18. Kemper, A.: Valuation of Network Effects in Software Markets: A Complex Networks Approach, p. 142. Springer (2009)
19. Wilensky, et al.: HubNet. http://ccl.northwestern.edu/netlogo/hubnet.html. Center for Connected Learning and Computer-Based Modeling, Northwestern University. Evanston, IL (1999)
20. Lam, C.Y., Lin, J., Sim, M.S., Tai, K.: Identifying Vulnerabilities in Critical Infrastructures by Network Analysis. Int. J. Crit. Infrastruct. **9**(3), 190–210 (2013)

21. Kizhakkedath, A., Tai, K., Sim, M.S., Tiong, R.L.K., Lin, J.: An agent-based modeling and evolutionary optimization approach for vulnerability analysis of critical infrastructure networks. In: AsiaSim 2013—Proceedings of the 13th International Conference on Systems Simulation, Singapore, Communications in Computer and Information Science, vol. 402, pp. 176–187. Springer (2013)
22. Tai, K., Kizhakkedath, A., Lin, J., Tiong, R.L.K., Sim, M.S.: Identifying Extreme Risks in Critical Infrastructure Interdependencies. In: ISNGI 2013—International Symposium for Next Generation Infrastructure, Wollongong, Australia (2013)

Introducing Cyber Security at the Design Stage of Public Infrastructures: A Procedure and Case Study

Sridhar Adepu and Aditya Mathur

Abstract Existing methodologies for the design of complex public infrastructure are effective in creating efficient systems such as for water treatment, electric power grid, and transportation. While such methodologies and the associated design tools account for potential component and subsystem failures, they generally ignore the cyber threats; such threats are now real. This paper presents a step towards a methodology that incorporates cyber security at an early stage in the design of complex systems. A novel graph theoretic mechanism, named Dynamic State Condition Graph, is proposed to capture the relationships among sensors and actuators in a cyber physical system and the functions that are affected when the state of an actuator changes. Through a case study on a modern and realistic testbed, it is shown that introducing security at an early stage will likely impact the design of the control software; it may also lead to additional hardware and/or software requirements, e.g., sensors, or secure control algorithms. Such impact on the system design promises to improve the resilience of a system to cyber attacks.

Keywords Cyber attacks · Cyber security · Cyber physical systems · Security by design · Dynamic state condition graph · SCADA · Water treatment

This work was supported by research grant 9013102373 from the Ministry of Defense and this work was supported by the National Research Foundation (NRF), Prime Minister's Office, Singapore, under its National Cybersecurity R&D Programme (Award No. NRF2014NCR-NCR001-40) and administered by the National Cybersecurity R&D Directorate.

S. Adepu (✉) · A. Mathur (✉)
iTrust, Center for Cyber Security Research, Singapore University of Technology
and Design, Singapore, Singapore
e-mail: sridhar_adepu@sutd.edu.sg

A. Mathur
e-mail: aditya_mathur@sutd.edu.sg

© Springer International Publishing Switzerland 2016
M.-A. Cardin et al. (eds.), *Complex Systems Design & Management Asia*,
Advances in Intelligent Systems and Computing 426,
DOI 10.1007/978-3-319-29643-2_6

75

Acronyms

CPS	Cyber physical system
DPIT	Differential pressure indicator and transmitter
DPSH	Differential pressure switch
LIT	Level indicator and transmitter
MV	Motorized valve
PLC	Programmable logic controller
PSH	Pressure switch
RO	Reverse osmosis unit
SCADA	Supervisory control and data acquisition
DSCG	State condition graph
SWaT	Secure water treatment
UF	Ultrafiltration unit
UV	Ultraviolet (dechlorinator)

1 Introduction

Cyber Physical Systems: A Cyber Physical System (CPS) consists of a physical process controlled by a computation and communications infrastructure. Typically, a CPS will have several Programmable Logic Controllers (PLCs) each with control software for computing control actions. Each PLC controls a portion of the entire process. The control actions are based on the current state of the system obtained through a network of sensors. When effected, and assuming the hardware effected is in working condition, the control action causes a desired change in the process state. For example, in a water treatment system, a PLC may start a pump to fill a tank with water to be sent through an ultrafiltration system. The pump must be stopped when the tank reaches a predetermined high level. The level of water in the tank is known to the PLC through a level sensor.

The PLCs in a CPS can be viewed as a system that transforms the state of the process as in Fig. 1. At any instant the PLCs receive data from sensors, compute control actions, and apply these actions to specific devices. Note that there are several potential attack points in a CPS. In this work only the communication links between sensors to PLC, denoted as a black blob in Fig. 1, are considered.

Response of a CPS under cyber attacks: The communications infrastructure of a CPS is often connected to an external network. Such connections render a CPS susceptible to cyber attacks. The presence of wireless communications among the CPS infrastructure, makes it even more vulnerable to cyber attacks. Such attacks could compromise the communications links between sensors and PLCs and among the PLCs. Once any communications link has been compromised, an attacker could use one of several strategies to send fake state data to one or more PLCs. Unless the defense mechanism of the attacked CPS is highly robust, such attacks could cause

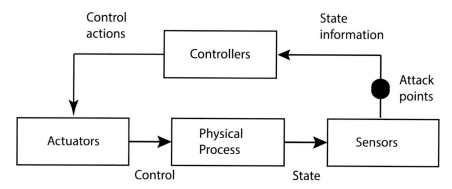

Fig. 1 CPS as a state transformer. In a water treatment system, actuators include pumps and motorised valves, while the sensors include level sensors, pH meters, chlorine sensors, and ORP (Oxidation Reduction Potential) meters. The *dark blob* indicates attack points considered in this work

an undesirable response from the CPS that may lead to system shutdown and/or device damage. Examples of such strategic attacks are given in Sect. 4.3. Thus, it becomes imperative for engineers to understand the response of a CPS to a variety of strategic cyber attacks and assess the robustness of its defense mechanism. An investigation like the one briefly described here is also critical in identifying errors in the control algorithms used by the PLCs though detection of such errors is not considered in this work.

Problem setting: It is assumed that a design consisting of various CPS components, and their interconnections, is available. For example, for a water treatment plant used in a case study, the physical subsystem of such a design would consist of pumps, tanks, valves, sensors, chemicals, etc., and the connecting pipes. The cyber component would consist of communications networks and various computing devices such as PLCs and other computing infrastructure. At this stage only the design of the physical system is available, and the control algorithms have not yet been coded. Prior to the actual construction of the CPS, and coding the control algorithms, it is desirable to know how would the system respond to cyber attacks. Thus, the problem can be stated briefly as follows.

(a) Using its design, determine how would a CPS respond to a set of potential cyber attacks and (b) how would the responses so obtained affect the design of the physical system and the control algorithms so as to improve its resilience to cyber attacks?

Contributions: (a) A scalable and automatable security-by-design procedure to understand the response of a Cyber Physical System to attacks on its communications infrastructure. (b) Dynamic State Condition Graph (D-SCG) as a formal modeling device for sensor-actuator constraints in a CPS.

Organization: The remainder of this work is organised as follows. Section 3 presents a step by step process for security by design of CPS. Section 4 presents a general CPS architecture, attacker models, and the DSCG. This section also contains examples to illustrate a procedure based on DSCGs for impact analysis of cyber attacks. Section 5 presents a case study to demonstrate how an DSCG-based procedure can be applied to analyze the defense mechanism of an operational water treatment system. Questions regarding the novelty, automation, and scalability of the proposed approach are discussed in Sect. 6. Related research and how it differs from that presented here is in Sect. 2. A summary, discussion, and next steps in this research appear in Sect. 7.

2 Related Work

A large body of work focusing on the modeling and analysis of CPS systems is available. Given that this work is concerned with using special kind of graphs to construct a formal design procedure, in this section works related to graphs in CPS is considered.

Topological vulnerability analysis: Jajodia et al. [1] proposed a detailed procedure for modeling cyber systems using attack graphs. Such graphs model practical vulnerabilities in distributed networked systems. While attack graphs model vulnerabilities, DSCGs do not. In fact DSCGs simply model conditions required to control a component in a CPS; vulnerabilities, if any, are discovered through an analysis procedure described in this paper. The attack graphs and DSCGs are similar in the sense that both model all paths through a system. Note that while DSCGs are specifically designed to model CPS, attack graphs are not.

Control flow integrity: Abadi et al. [2] propose a control-flow integrity (CFI) approach to mitigating cyber attacks on software. They claim that CFI "…can prevent such attacks from arbitrarily controlling program behavior." CFI differs from DSCGS in many ways. First, CFI approach targets only software whereas DSCG target both hardware and software. Second, CFI is aimed at ensuring that malware does not affect the behavior of software. DSCGs do not focus on malware. Third, as mentioned above, DSCGs aim at modeling the behavior of controllers and hence obtain only conditions that must be true for a control action which is not the focus of CFI.

Other modeling approaches: Chen et al. [3] have proposed argument graphs as a means to capture the workflow in a CPS. The graphs are intended to assess a system in the presence of an attacker. The graphs are formed based on information in the workflow such as use case or state, physical system topology such as network type, and attacker model such as order to interrupt, power supply, physical tampering, network connection, Denial of service, etc. Vertices in these graphs contain the information corresponding to certain classes of security related information; they do not capture conditions for successful control actions. Instead, the graphs assume that the existence of flow implies a secure system.

Argument graphs are considered corresponding to each use case. When all use cases are considered, the graphs become unwieldy and difficult to analyze. The DSCG graphs are drawn from the conditions used in a CPS or its design, and hence enable the analyzes of the damage or other effects when properties of the system are altered as may happen when an attacker enters a system via cyber or physical means.

Would an extension of argument graphs make them look and function like DSCGs? The answer is in the negative because the inputs to argument graphs and those in DSCGs are different. The vertices in DSCGs are physical devices and edges are conditions to initiate actions. In argument graphs the vertices may contain information about sub-steps in the use case, attacks and network related devices. Attack graphs are mostly helpful to find out the security level of the system in "Layer 1" and "Layer 2" in the SWaT system whereas DSCGs are generated based on the level 0 physical devices and their control dependency on one another.

Typed graphs [4] and Baysian defense graphs: [5] are a few other important contributions to the modeling of cyber attacks. However, once again, these differ from DSCGs in aspects already mentioned above.

Robust CPS: Here exists literature on the design of robust CPS [6–8]. These works focus on attack modeling, the design of controllers and monitors for secure CPS. In this paper, attack models are borrowed from Cardenas' work [9] as this is most closely related to cyber attacks in a CPS. Many other works model attacks specifically on control systems and are abstract in nature. DSCGs allow one to model attacks in a very practically visible way through the use of design diagrams.

Robust CPS: There is a large body of literature on failure mode and effects analysis of physical systems [10, 11]. These methods, also referred to as FMEA, are aimed at assessing the reliability of a system in the presence of failure of its components. Indeed, FMEA is a useful technique for assessing the reliability of the physical portion of a CPS. However, the presence of control software renders FMEA difficult and sometimes impractical to use due to software complexity. In fact the analysis procedure presented here using DSCGs can be considered as a new variant of FMEA and more appropriately referred to as SMEA: Security Mode Effects and Analysis. Note that, though not discussed here, failure can also be modelled using DSCGs via minor modifications to the graph semantics.

3   Security by Design: A Process

CPS design begins with a clear understanding of the requirements. These requirements indicate the expected, behavior of the CPS to realise an end objective such as deliver electric power to customers or produce clean water. How failure of system components needs to be managed is either a part of the requirements or added using techniques such as failure mode and effects analysis. Experience with the design of CPS suggests that while component failure scenarios are included in the CPS design, cyber attack scenarios are not. While some cyber attacks might lead to failure-like conditions, other strategic attacks might not. Even though a cyber-attack might lead to a failure-like scenario, the malicious intent of an attacker significantly increases the occurrence probability of an otherwise low-probability

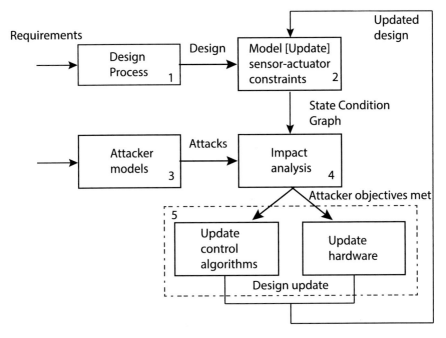

Fig. 2 An iterative design process to strengthen hardware and control algorithms to mitigate the effect of cyber attacks

failure scenario. This makes it important to include a security analysis component in the design process described below with reference to Fig. 2.

Step 1: CPS design: Given the CPS requirements, a design team creates the design of a physical process. This design is in the form of an engineering schematic, supported by textual explanation such as in [12]. The schematic shows the various physical components, their interconnections, and possibly mechanical and electrical specifications. It is assumed here that the computer programs for controlling the CPS are written after at least an initial physical design is available. Certainly, in some cases, code could be borrowed from a previous similar design.

Step 2: CPS model: The design of the physical system is then used to derive constraints that must be satisfied for each actuator in the system to be in a given state. Grouped together, these constraints lead to an DSCG as explained in Sect. 4.2.

Step 3: Attacker model: An attacker is modeled in this step in terms of objectives and the attack means. As explained in Sect. 4.3, each attacker model may lead to more than one attack.

Step 4: Impact analysis: As described in Sect. 4.4, attacks generated using attacker models are used as input to the DSCG. The ensuing analysis leads to the response of the CPS to different attacks and whether or not the attacker objectives are met. A given attacker model may generate an unusually large number of attacks. Thus, the choice of which attacks to select for impact analysis becomes an issue.

Later we explain how a DSCG aids in making such a choice based on economic arguments.

Step 5: **Design update**: Impact analysis reveals whether the attacker objective can be met or not. The impact analysis could be carried out with or without the PLC control software in place. If the PLC software is in the model, the impact analyses reveals any weaknesses of the software, such as a missing check that might otherwise reveal a cyber attack or a component failure. An indication that the attacker objective can be met implies weaknesses in system defense assuming the current design. It also offers clues as to what hardware and software defense are needed to reduce the chances of the attack being successful. Any changes made in this step in hardware design requires updates to the model (Step 2) and re-execution of steps 4 and 5. The loop consisting of steps 2, 4 and 5, ought to be executed in the design process as long as the impact analysis reveals weaknesses in the design based on the attacker models. This process is illustrated in Sect. 4.5.

4 Modeling a CPS

The first step in the proposed procedure is to construct a suitable model of a CPS. While a model using tools such as Simulink [13] or Labview [14] are certainly aids in the procedure described here (especially in Step 4 in Fig. 2), the procedure described here is a complementary aid. A general architecture of a CPS and the modeling procedure based on this architecture, are described next.

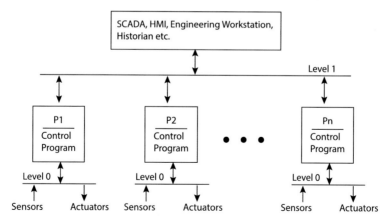

Fig. 3 Architecture of the control portion of a CPS. P1, P2, ..., Pn denote PLCs. Each PLC communicates with its sensors and actuators through a local network at Level 0. PLCs communicate among themselves via another network at Level 1. Communication with SCADA and other computers is via a Level 3 network not shown here. Note that the actuators, e.g., a pump, also have sensors to indicate their condition

4.1   Structure of a CPS

CPS, such as power grid and water treatment systems, consist of a distributed supervisory control system. The control system itself is a collection of PLCs each controlling a specific portion of the CPS. This architecture is exhibited in Fig. 3. As shown, each PLC communicates with a set of sensors and actuators via a local network. This network is considered to be at Level 0 and is also referred to as the field-bus network [15]. The PLCs communicate with each other using the Level 1 network. Such a layered network structure is in accordance with the prevailing practice for industrial control systems [16].

As in Fig. 3, each PLC is responsible for the control of a set of actuators. The control actions are computed based on data received from a set of sensors local to it as well as data obtained from other PLCs. Data from sensors local to other PLCs is obtained via the Level 1 network through a sequence of message request and response.

Each PLC contains a control program that receives data, computes control actions and applies these to the actuators it controls. Computation of control actions is based on a condition evaluated using data received from the sensors. This could be a simple condition involving data from one sensor, or a compound condition involving data from multiple sensors some of which might be communicating with other PLCs. As described next, it is these conditions that are captured in the form of an annotated dependency graph to create a model for a CPS.

4.2   Dynamic State Condition Graphs

A Dynamic State Condition Graph (DSCG), is a pair (N, E), where N is a finite set of labeled nodes and E a finite set of (possibly) labeled directed edges. Three types of nodes are considered. A state-node, referred to as *s-node*, denotes the state of an actuator such as a pump, a tank, a generator, or a tap changing transformer. For an actuator with k states, there are k nodes in the corresponding DSCG, one denoting each state. A component node, referred to as a *c-node*, denotes any component of a CPS that could be in any of two or more states. An operator-node, referred to as *o-node*, denotes a logical operator such as a logical and ($\wedge$), logical or ($\vee$), and logical not ($\neg$).

A labeled edge is a triple (n_1, l, n_2), where n_1 and n_2 denote, respectively, c-node and s-node, and l the state of the component denoted by n_1. n_2 denotes the state of an actuator. l could be specified as a condition, such as *pressure less than 3 Bar*, or as a discrete state, such as `C(losed)`.

Each s-node is labeled with one or more functions of time, and probably some other process parameters. Each function denotes a property of a physical component of the system being modeled. These properties are affected by a change in system state. While the s-nodes represent sensors and actuators of a CPS being modeled, it is these time-dependent functions that model the system dynamics and the name DSCG. In this case study, the functions representing product and component properties, were not used. However, these function are essential when a DSCG is used in creating a realistic simulation of the system.

For convenience, a DSCG is usually presented as a collection of sub-graphs, each corresponding to one or more components, or even the state of a component, of the CPS. Given the distributed and connected nature of a CPS control software, these subgraphs are connected. For example, one subgraph might indicate conditions for turning a pump ON. Another subgraph might use the ON state of the pump as a condition to change the state of another component.

Example 1 Consider a system S, as in Fig. 4a, consisting of the following components: a pump, a valve, and two water tanks. PLC 1 controls the pump while PLC 2 controls the valve. Level sensors at each tank communicate with the PLCs as shown. Each tank can be in any of four states: LowLow (LL), Low (L), High (H), and HighHigh (HH). The pump has two states: ON and OFF. The valve that connects pump to Tank B can be in one of two states: O(pen) and C(losed).

Now suppose that the design of S requires the following conditions to govern the pump. The pump is started, i.e., its state changed from OFF to ON, when Tank A is not in state LowLow, Tank B is not in state HighHigh, and the valve is open (O). The pump state is changed from ON to OFF whenever Tank B is in the HighHigh state or Tank A is in LowLow state. Figure 4 shows a partial DSCG for S that captures the conditions that govern the pump operation. This partial DSCG has two *s-nodes* labeled PON and POFF, two *c-nodes* labeled Tank A and Tank B, and two *op-nodes* labeled ∧ and ∨.

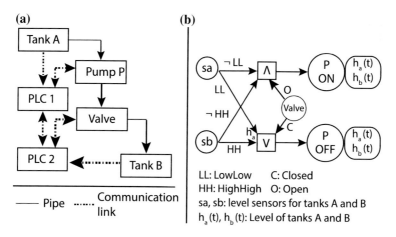

Fig. 4 a A subsystem of a CPS S consisting of two tanks, a pump, a valve, and two PLCs controlling the pump and the valve. **b** A portion of the DSCG for S depicting the conditions that govern pump operation and the time-dependent properties of the affected components. Conditions that govern the state change of the valve are not shown here

The state of pump P effects the water level in tanks A and B. Thus, when P is in state ON, the water level in tank A, $h_a(t)$, reduces while that in tank B, $h_b(t)$ increases. Functions $h_a(t)$ and $h_b(t)$ depend on the mechanical properties of the pump and dimensions of the two tanks. The discrete state assigned to a tank depends on the values of the corresponding height function $h(t)$. ∎

4.3 Attacker Models

For a CPS, one possible attacker model is a pair (T, O), where T is an attack type to realise objective O. The attack type could be of any of the types proposed earlier such as in [6, 8, 9, 17]. More complex attack types are possible. The objective is specified as a statement. For example, *"Damage generator A in a power grid,"* or *"Damage pump P302 in a water treatment network."* A *cyber* attack is a sequence of actions, a procedure, initiated by the attacker where each action is initiated via a cyber component, such as a wireless link or a SCADA computer. When the attack is via a physical component, such as an explicit damage to a pump or the physical removal of a circuit breaker, it is considered a *physical* attack. The actions in an attack are selected and sequenced so as to model the attack type T and realise the objective O. Whether or not the attempted action sequence will realise the attacker objective depends primarily on the defense mechanism used in the CPS.

Example 2 Consider the following attacker objective and attack type for the system in Fig. 4: *Cause Tank B to overflow* using a *deception attack* [18]. The attacker uses the following procedure to achieve the objective.

(1) **Enter and capture**: Identify the wireless communication links and capture the link from Tank B to PLC 2.
(2) **Wait and listen**: Listen to the data transferred across the links. Wait until Pump B is ON, the valve is C(losed), and Tank B is close to entering the HH state, say, when it is in state H.
(3) **Deceive**: Regardless of the data input from the Tank B level sensor, send to PLC 2 a value that corresponds to H.
(4) **Wait and listen**: Continue monitoring the Tank B level sensor until a few minutes after it outputs a value that corresponds to HH. An overflow will occur if the pump has not been shut sometime after Tank B moves to HH. The exact time when the overflow occurs depends on the excess capacity in Tank B beyond that needed in HH.
(5) **Exit**: Exit from the system when satisfied that the overflow has occurred. ∎

4.4 Impact Analysis

Each possible action in the attack needs to be analyzed for its possible impact. This analysis aids in identifying possible weaknesses in the current defense mechanism, and hence in making it more robust. The complete DSCG graph, or a simulation based on it, is traversed to determine the potential impact of each action; the actual impact can be determined by implementing the attack on a realistic testbed as done in this case study. It is indeed possible that while the attacker objective may or may not be realized through the proposed actions, undesirable side effects might. It is best to perform this analysis from a pessimistic view. For example, instead of assuming that a given action, such as capture of a wireless link, is infeasible, it might be wise to assume that it is. Note that the impact of a cyber attack depends on the state of the system at the time of launch. However, this aspect is not considered in this paper.

Example 3 A brief sketch of impact analysis is presented next based on the actions described in Example 2. The focus here is on action "Deceive" as the impact of the other actions is relatively easy to determine.

At the time the deception action is initiated, the inputs to the OR ($\wedge$) node in Fig. 4b are: $\neg$ LL, C(losed), and $\neg$ HH. In fact this is the correct state of the SWaT. However, depending on the rate of flow, Tank B will soon be in HH state while the input to PLC 2 will remain at H as determined by the attacker. At this time the deception action causes PLC 2 to incorrectly assume the state of the system; specifically the state of Tank B. This state divergence, i.e., the difference between the actual and the computed system states, remains until the attack is detected and the system reset.

Moving ahead, the incorrect state assumption by PLC 2 causes Pump B to remain ON as neither of the two conditions at the OR node is true. Over time, and unless there is an effective defense mechanism, Tank B will overflow.

The analysis now must continue with the remainder of the DSCG for S. Overflow of a tank might cause inconvenience or wasted water. It might also lead to more serious scenarios such a electrical short circuit and its impact on the control devices. ∎

4.5 Design Update

Impact analysis will likely expose weaknesses in the CPS defense mechanism. In turn, CPS designers could then decide whether to remove the exposed weaknesses, or to let them remain perhaps because of the low probability of success of the attack that exposed the weakness, being successful, or due to reasons of economy.

Example 4 The analysis in Example 3 reveals a potential weakness in the defense against an attack on the Tank B level sensor. There are several defenses against this attack. A hardware defense is to create a mechanical interlock that ensures (a) the pump is shut off automatically when Tank B is in state HH, and (b) an alarm is raised at the SCADA as well as physically near Tank B.

A software defense is also possible, though is more complex than the above hardware defense. PLC 2 could use a model that allows it to compute the water level in Tank B. The computed level could be compared with that received from the sensor. Any significant discrepancy is a cause for alarm. Several other possibilities exist, not discussed here, for enhancing the system defense against deception attacks. ■

5 Case Study

The design and analysis approach described above was used to analyze the defense mechanism of a Secure Water Treatment (SWaT) testbed. While the proposed procedure is intended to be applied during the CPS design process, in the case study reported here the procedure was applied on an operational system. As is often the case, SWaT was designed and built for correct operation. While the control algorithms in SWaT do account for component failures, they are not designed to detect and defend against cyber attacks. This aspect of SWaT makes it a useful subject to study the effectiveness of the DSCG-based modeling and analysis procedure.

5.1 Architecture of SWaT

SWaT is a testbed for water treatment. In a small footprint producing 5 gallons/h of filtered water, the testbed mimics a large modern water treatment plant found in cities. It is used to investigate response to cyber attacks and experiment with novel designs of physics-based and other defense mechanisms. As shown in Fig. 5, SWaT consists of six stages labeled P1 through P6. Each stage is controlled by its own set of dual PLCs, one serving as a primary and the other as a backup in case of any failure of the primary.

Communications: Each PLC obtains data from sensors associated with the corresponding stage, and controls pumps and valves in its domain. Turning the pumps ON, or opening a valve, causes water to flow either into or out of a tank. Level sensors in each tank inform the PLCs when to turn a pump ON or OFF. Several other sensors are available to check the physical and chemical properties of water flowing through the six stages. PLCs communicate with each other through a separate network. Communications among sensors, actuators, and PLCs can be via either wired or wireless links; manual switches switch between the wired and wireless modes.

Stages in SWaT: Stage P1 controls the inflow of water to be treated by opening or closing a valve (not shown) that connects the inlet pipe to the raw water tank.

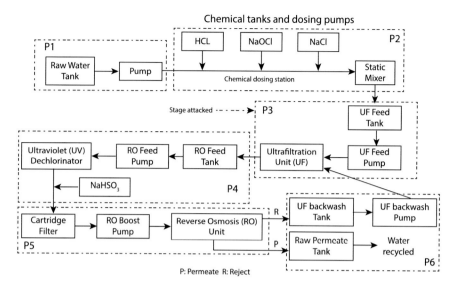

Fig. 5 Physical water treatment process in SWaT and attack point used in the case study. P1 though P6 indicate the six stages in the treatment process. The attack reported in this paper is on stage P3. Each stage is controlled by its own PLC connected to sensors and actuators. PLCs communicate among themselves and the SCADA computer via separate networks

Water from the raw water tank is pumped via a chemical dosing (stage P2) station to another UF Feed water tank in the stage P3.

In stage P3, a UF feed pump sends the water via UF membrane to RO feed water tank in stage P4. Here an RO feed pump sends the water through an ultraviolet dechlorination unit controlled by a PLC in stage P4. This step is necessary to remove any free chlorine from the water prior to passing it through the reverse osmosis unit in stage P5. Sodium bisulphate ($NaHSO_3$) can be added in stage P4 to control the ORP.

In stage P5 the dechlorinated water is passed through a 2-stage RO filtration unit. The filtered water from the RO unit is stored in the permeate tank and the reject in the UF backwash tank. Stage P6 controls the cleaning of the membranes in the UF unit by turning on or off the UF backwash pump. The backwash cycle is initiated automatically once every 30 min and takes less than a minute to complete. Differential pressure sensors in stage P3 measure the pressure drop across the UF unit. A backwash cycle is also initiated If the pressure drop exceeds 0.4 bar indicating that the membranes need immediate cleaning. A differential pressure meter installed in stage P3 is used by PLC 3 to obtain the pressure drop.

Cyber attack points in SWaT: In this case study the wireless links between sensors and the corresponding PLCs are considered as the attack points. A pessimistic approach is taken implying that all wireless links are assumed to be vulnerable to cyber attacks. Initial experiments, not described here, revealed that indeed, wireless communications in SWaT are vulnerable. In this study various level transmitters were considered as attack points.

5.2 Attacker Models

Attacker models are needed to understand the response of SWaT to malicious attacks. Table 1 lists four attack types and, for each type, one attacker objective. The attack types used here have been proposed by Cardenas et al. [9]. Note that a variety of attacks can be generated using each attacker model and attack type. Only one attack of each attack type is reported in this case study as an illustration of the design analysis procedure.

5.3 Modeling SWaT

This case study was performed on an operational system. While the complete design of SWaT is available to the authors, the DSCG model was created using actual ladder logic and structured text code [19] that resides in the six PLCs. The choice of PLC code, instead of using the design, was motivated by the desire to obtain an accurate model of SWaT.

SWAT model in terms of DSCG subgraphs consists of total of 12 sub-graphs. The sub-graphs are connected through links across the PLCs. Conditions governing the control of each pump and each motorized valve were modeled. Due to space limitations the DSCGs are not shown here.

5.4 Choice of Attacks and Impact Analysis

Detailed impact analysis was conducted using the 12 DSCGs developed from the PLC code. Each DSCG corresponds to conditions to change the state of an actuator in SWaT. Four cyber attacks were selected, one for each attack type in Table 1. Here only one implementation of the surge attack is described. The objective of the attack was to damage the UF unit. Assuming that an attacker knows the mechanical

Table 1 Attacker models used in the case study

Attack type	Attacker objective
Bias	Disturb chemical dosing
Covert	Affect dechlorination
Replay	Affect water quality
Surge	Damage the ultrafiltration unit

and chemical properties of the UF unit, and the overall structure of SWaT, attacks to damage the UF unit can be derived. These attacks would likely be successful when appropriate defense mechanisms do not exist.

The attacks in this case study were launched from the SCADA computer. These could also be launched in SWaT via the wireless network that connects the sensors to the PLCs. The actions needed to damage the UF unit are in Table 2 and numbered 1 through 4. The second from left column in the table lists attacker actions to achieve the desired objective. These actions are derived from the subgraph in DSCG that corresponds to pump P301. The rightmost column lists the consequence of each attacker action derived also from DSCG. Note that the attacker actions cause a discrepancy between the actual system state and the state known to PLC 3. The consequence of each attacker action, expressed in terms of conditions evaluated by PLC 3, are derived using this DSCG. The eventual impact on the UF unit is not derived using DSCG; it is derived through a series of arguments, not mentioned here, based on the mechanical properties of the UF membranes.

5.5 Impact Analysis Summary

A summary of all four types of attacks considered and their potential impact on SWaT appears in Table 3. Note that due to lack of hardware and software defense mechanisms, SWaT components such as UF and RO are likely to be damaged if these attacks were implemented and sustained for a long period. This claim of

Table 2 Impact analysis using a DSCG; attacker objective: damage the ultrafiltration unit

Attack	Actions	Consequence
1	Spoof messages going to PLC 3 by compromising the wireless link from the sensors	Attacker can send false data to PLC 3
2	Set the high pressure sensor PSH-301 to 2.0 Bar	System state: PSH301 > 2.5 Bar In PLC: PSH301 < 2.5 Bar Hence, in the absence of the attack, P301 should be turned OFF, but as the PLC has the incorrect state information, it does not turn P301 OFF
3	Set the differential pressure switch DPSH-301 to 0.3 Bar	System state: DPSH301 > 0.5 Bar In PLC: DPSH301 < 0.5 Bar Hence, in the absence of the attack, P301 should be turned OFF, but as the PLC has the incorrect state information, it does not turn P301 OFF
4	Set the differential pressure indicator DPIT-301 to 0.3 Bar	System state: DPIT301 > 0.4 Bar In PLC: DPIT301 < 0.4 Bar

Impact on SWaT: UF does not enter immediate backwash cycle; UF deterioration accelerated; UF is likely to be damaged if the attack persists for sufficient time. The time to damage the UF will depend on the incoming water quality and the properties of the membranes in the UF unit

Table 3 Summary of impact analysis on SWaT

Attack type	DSCG used	Outcome	Damage
Bias	p2_off	Dosing does not get activated to change the water properties	Water produced does not maintain desired chemical properties
Covert	p4_on	Water dechlorination does not take place for 10 min	Increased chances of damage to the RO unit
Replay	p5_on	Impure water gets into the RO unit permeate tank	No hardware damage
Surge	p3_off	Ultrafiltration unit damage accelerated due to delay in backwash	Increased chances of UF damage

p2, p3, p4, and p5 correspond to, respectively, three dosing pumps, pump P301, pump P401, and pump P501. Thus, p2_off, p3_off, p4_on, and p5_on refer to their respective DSCGs

damage is based on the mechanical properties of the membranes in the UF unit. Also, the claim of damage is being made cautiously as regular physical checks of water quality could enable attack detection prior to component damage.

5.6 Design Update

Based on the impact analysis described above, a detailed design of the defense mechanisms ought to be considered. Such a design has not been attempted so far. Nevertheless, a few potential hardware and software defense mechanisms are considered next.

Improving the security of the wireless connections is an obvious defense against all spoofing attacks. In the present context, more interesting defense mechanisms related to actions 2 through 4 in Table 2 are considered. These actions prevent urgent cleaning of the UF unit. However, the regular 30 min cleaning cycle will still be active. If the attack happens soon after a regular cleaning cycle, then the UF will be operating, with clogged membranes, for at most about 29 min. Thus, depending on the quality of the incoming water, and characteristics of the UF membranes, damage will likely occur.

One defense against the above attack is to install one or more water quality meters in the pipe that carries water from UF output to the RO feed tank. While several such a quality meters are available in SWaT but not immediately following the UF unit. Installation of additional water quality sensors will require the PLC code to be updated. Doing so will also update the corresponding DSCGs for any further impact analysis.

Another defense against the above, as well as other cyber attacks, is to have an *independent* network of sensors [20] that regularly check the health of the system, and especially of the critical system components such as the UF and RO units. These sensors do not communicate with the PLCs but have a *one way* communication with the SCADA system. The sensors continuously check against violation of water quality constraints and raise appropriate alarms. The independence of the sensor network is crucial in this mechanism to be effective against cyber intrusions.

6   Novelty, Automation, and Scalability

Novelty: Given that a large number of approaches exist for modeling CPS, the novelty of the DSCG-based approach to impact analysis needs to be addressed. Approaches known to the authors and cited later in Sect. 2 do not explicitly include (a) the cyber and physical components of a CPS and (b) the conditions that affect the state of an actuator. Such an inclusion in a DSCG adds significant value in the early analysis of a CPS design. First, by creating a graph adjacency matrix from the system DSCG, one can easily and automatically, determine how many and which components of a system could be impacted when a given sensor is attacked. This information is highly valuable both for an attacker and the designer. It allows the attacker to design attacks for maximum system damage or disruption while offering valuable hints to the designer as to where to spend the effort in placing hardware/software defense mechanisms. Thus, for a designer this analysis offers economic arguments in favor or against adding specific security hardware/software. Though straightforward, this analysis is not possible using models such as Petri net [21] or graph based [1, 22]. Another straightforward outcome of a DSCG is the path condition for different attacks. Thus, by traversing a DSCG from a specific sensor node to any other node, one can determine what conditions are necessary for an attack to be successful or not successful. Note that either of the analyses mentioned above are valid for both cyber and physical attacks.

Automation: As in Fig. 2, a DSCG is created from a CPS design. This cannot be done without redesigning existing system design tools [23, 24]. However, given a version of a design schematic that labels various components, it is a matter of software design to map any CPS design to a DSCG. Note that a schematic will likely not include time functions that describe the property of each system component, e.g., electric generator, or a partial product, e.g., water flow rate out of a tank, on the state of one or more actuator states. Perhaps with the aid of a database of well known physical properties of components, the task of associating product and component state and product property functions could also be automated. While the difficulty of automating the entire DSCG construction will be best gauged when this task has been completed at least once, no significant technical hurdles seem to be on its way.

Scalability: The scalability of a DSCG-based design analysis approach depends on at least two factors: number of sensors and actuators in a CPS under design, and the selection of cyber and physical attacks. The number of nodes in a DSCG is linear in the number of sensors and actuators. Given a CPS design with N sensors and M actuators, the total maximum number of subgraphs in a DSCG is kM, where constant k depends on the number of discrete states of each actuator. Thus, the number of subgraphs grows linearly with actuators. The number of actual nodes in each sub-graph depends on the number of components controlled by each actuator. Again, this is linear in the number of system components. The adjacency matrix is of size C^2, where $C = N + M$. Thus, for a system with, say, $C = 3000$, the adjacency matrix will have a total of 9 million entries. Fast algorithms for transitive closure [25] could be used to rapidly perform reachability analysis.

Given the attack model in this paper, the potential number of attacks is exorbitant when a component state is represented by a real number. However, in practice, continuous state space is often reduced drastically through discretization and engineering design so that only a few useful and critical states are considered during analysis. Doing so reduces the number of potential cyber attacks to grow linearly in the number of cyber components in a CPS and grows similarly in the number of physical components. Further reduction in the number of attacks to be used during the analysis phase is possible using arguments based on the economics of the system. A discussion on these aspects of attack space reduction is beyond the scope of this paper.

7 Summary, Discussion, Next Steps

A procedure to model a CPS at the design and operational stage is proposed. The procedure is based on Dynamic State Condition Graphs that capture the conditions used by control algorithms to change the state of individual CPS components. The proposed procedure has been applied to study the vulnerabilities in the software and hardware design of a modern and realistic water treatment system. The analysis revealed several weaknesses in the system design. While the system was designed to function correctly, security was a minor factor in the design. Thus, the DSCG-based procedure helped in identifying various weaknesses and hinting at software and hardware means for their removal.

The use of DSCGs presents a simple and practically usable procedure to assess the defenses of a CPS. Simplicity, and hence its ease of use, is a key characteristic of the procedure. The case study presented in this paper offers a glimpse into how the notion of "Security by Design" can be realised in practice. The approach is realistic and does not rely on any form of abstraction such as that found in linear control flow models of systems [6, 26, 27]. Further, the graphical nature allows partial automation to understand how an attack progresses through a CPS.

The analysis presented in the case study in this paper was done manually. The graphs, not shown in this paper, are constructed using a Python program but the impact analysis was done manually. The analysis procedure needs some automation for it to be applicable in the design of realistic systems. However, doing so requires a clear understanding of component semantics such as when does a component fail. DSCGs could become an even more powerful tool once they are enhanced with physical operational constraints of each device included in the model.

Acknowledgments Thanks to: Nils Tippenhauer, Daniel Daniele Antonioli for demonstrating the feasibility of attacking wireless links between sensors and PLCs in SWaT; Kaung Myat Aung for assisting in the validation of constraints in DSCG; and Ivan Lee, Mark Goh, and Angie Huang for their constant assistance without which this research would not be possible.

References

1. Jajodia, S., Noel, S.: Advanced cyber attack modeling, analysis, and visualization. Technical Report AFRL-RI-RS-TR-2010-078. Final Technical Report, George Mason University (2010)
2. Abadi, M., Budiu, M., Erlingsson, U., Ligatti, J.: Control-flow integrity principles, implementations, and applications. ACM Trans. Inf. Syst. Secur. **13**(1), 4:1–4:40 (2009)
3. Chen, B., Kalbarczyk, Z., Nicol, D.M., Sanders, W.H., Tan, R., Temple, W.G., Tippenhauer, N.O., Vu, A.H., Yau, D.K.: Go with the flow: toward workflow-oriented security assessment. In: Proceedings of the 2013 Workshop on New Security Paradigms Workshop, NSPW '13, pp. 65–76 (2013)
4. Bhave, A., Krogh, B., Garlan, D., Schmerl, B.: View consistency in architectures for cyber-physical systems. In: Proceedings of 2nd ACM/IEEE International Conference on Cyber-Physical Systems (2011)
5. Sommestad, T., Ekstedt, M., Johnson, P.: Cyber security risks assessment with Bayesian Defense graphs and architectural models. In: 42nd Hawaii International Conference on System Sciences, pp. 1–20 (2009)
6. Kwon, C., Liu, W., Hwang, I.: Security analysis for cyber-physical systems against stealthy deception attacks. In: American Control Conference (ACC), 2013, pp. 3344–3349 (2013)
7. Pasqualetti, F., Dorfler, F., Bullo, F.: Attack detection and identification in cyber-physical systems. IEEE Trans. Autom. Control **58**(11), 2715–2729 (2013)
8. Wasicek, A., Derler, P., Lee, E.: Aspect-oriented modeling of attacks in automotive cyber-physical systems. In: Design Automation Conference (DAC), 2014 51st ACM/EDAC/IEEE, pp. 1–6 (2014)
9. Cárdenas, A.A., Amin, S., Lin, Z.-S., Huang, Y.-L., Huang, C.-Y., Sastry, S.: Attacks against process control systems: risk assessment, detection, and response. In: ACM Symposium on Information, Computer and Communication Security (2011)
10. Kara-Zaitri, C., Keller, A., Barody, I., Fleming, P.: An improved fmea methodology. In: Reliability and Maintainability Symposium, 1991. Proceedings, Annual, pp. 248–252 (1991)
11. Li, J., Xuan, C., Shao, B., Ji, H., Ren, C.: A new connected device-based failure mode and effects analysis model. In: 2014 IEEE International Conference on Service Operations and Logistics, and Informatics (SOLI), pp. 1–5 (2014)
12. Young, W., Stamp, J., Dillinger, J.: Communication vulnerabilities and mitigations in wind power SCADA systems. In: American Wind Energy Association WINDPOWER Conference Austin, Texas, pp. 1–15 (2003)
13. http://www.mathworks.com/products/simulink/
14. http://www.ni.com/labview/
15. Stouffer, K., Scarfone, J.F.K.: Guide to Industrial Control Systems (ICS) Security (2011)
16. Galloway, B., Hancke, G.: Introduction to industrial control networks. IEEE Commun. Surv. Tutorials **15**(2), 860–880 (2013)
17. Amin, S., Càrdenas, A., Sastry, S.S.: Safe and secure networked control systems under denial-of-service attacks. In: Hybrid Systems: Computation and Control. Proceedings of 12th International Conference (HSCC), LNCS, vol. 5469, pp. 31–45. Springer (2009)
18. Amin, S., Litrico, X., Sastry, S., Bayen, A.: Cyber security of water SCADA systems; Part I: analysis and experimentation of stealthy deception attacks. IEEE Trans. Control Syst. Technol. **21**(5), 1963–1970 (2013)

19. Allen-Bradley: Logix5000 Controllers Structured Text, Programming Manual, Publication 1756-PM007D-EN-P, Rockwell Automation (2012)
20. Sabaliauskaite, G., Mathur, A.P.: Intelligent checkers to improve attack detection in cyber physical systems. In: Proceedings of the 2nd IEEE International Workshop on Cyber Security and Privacy (CSP 2013), International Conference on Cyber-Enabled Distributed Computing and Knowledge Discovery (CyberC 2013) Beijing, PRC (in press) (2013)
21. Chen, T., Sanchez-Aarnoutse, J., Buford, J.: Petri net modeling of cyber-physical attacks on smart grid. IEEE Trans. Smart Grid **2**(4), 741–749 (2011)
22. Bondavalli, A., Lollini, P., Montecchi, L.: Graphical formalisms for modeling critical infrastructures. In: Critical Infrastructure Security: Assessment, Prevention, Detection, Response. WIT Press Royal (2011)
23. Power plant design
24. Siemens: Programming with STEP 7 Manual, 05/2010, a5e02789666-01
25. Lidl, R., Pilz, G.: Applied Abstract Algebra, 2nd edn. Springer (1998)
26. Mo, Y., Hespanha, J., Sinopoli, B.: Robust detection in the presence of integrity attacks. In: American Control Conference (ACC), 2012, pp. 3541–3546 (2012)
27. Pasqualetti, F., Dorfler, F., Bullo, F.: Attack detection and identification in Cyber-Physical Systems, models and fundamental limitations. IEEE Trans. Autom. Control **58**(11), 2715–2729 (2013)

Ontology for Weather Observation System

Yuki Onozuka, Makoto Ioki and Seiko Shirasaka

Abstract Why do humans develop systems? The answer is clear: to realize system capabilities by utilizing these systems. Unless a system realizes the intended capability, it is useless, even if the system itself is perfect. Every system needs something to be realized. The problem is that we do not know how to deal with unexpected problems that prevent the realization of the system capability. Of course, many studies have been conducted on how to develop robust, fault tolerant systems, but there is no existing research on how to determine the system's resilience in relation to its capability. The system capability can be realized by developing a resilient system. In this study, we developed a system ontology especially for a weather observation system. This ontology can describe the entire system by applying DODAF 2.0 and an enabler relationship. It allows us to determine all of the system entities and relationships. In addition, the ontology shows the system's resilience by considering the realization of the system capability. This ontology was applied to a weather observation system to verify its effectiveness.

1 Introduction

1.1 Motivation

In the past, space has been utilized as a sanctuary because its environmental characteristics are completely different from those of the land, sea, and air. Recently, the space infrastructure situation has changed. As technology has advanced, humans have developed and launched weather and GPS satellites. The importance of space infrastructure has gained recognition all over the world. For example, we usually

Y. Onozuka (✉) · M. Ioki · S. Shirasaka
Graduate School of System Design and Management, Keio University,
4-1-1 Hiyoshi, Kohoku-Ku, Yokohama, Kanagawa, Japan
e-mail: y-onozuka@z5.keio.jp

© Springer International Publishing Switzerland 2016
M.-A. Cardin et al. (eds.), *Complex Systems Design & Management Asia*,
Advances in Intelligent Systems and Computing 426,
DOI 10.1007/978-3-319-29643-2_7

check the weather forecast before going outside and often utilize a map in order to efficiently reach our destination. Space infrastructure has become part of our daily life. What if someday we cannot use this infrastructure? What if our country's satellites are intentionally damaged by another country? In the past, some countries have succeeded in destroying satellites, which means today's space infrastructure could potentially be destroyed, either intentionally or by accident. Furthermore, a large amount of space debris is in orbit. The rising population of space debris increases the potential danger of accidental collisions [1]. Such collisions could eliminate the benefit derived from space infrastructure. However, we can develop a plan to avoid such a catastrophic situation.

For example, the United States has two kinds of counterplans to any threat to its space infrastructure. These represent a multi-layered approach to deterring attacks on its space capabilities and enhancing the resilience of these measures [2]. Japan also tries to enhance the resilience of its space infrastructure to deal with uncertain situations [3]. This resilience is defined as follows [4].

Resilience is the ability of an architecture to support the functions necessary for mission success in spite of hostile action or adverse conditions.

Based on this definition, a government can enhance the resilience by changing the space architecture design from intensive to dispersive [5]. However, in this research, we propose to enhance the resilience by developing an ontology that makes it possible to see the whole system. In other words, we develop an ontology to show the whole system and visualize how to maintain its capabilities.

In 2013, the Japanese government published its "Basic Plan on Space Policy," which provides a very important statement on its space policy [3]. Future decisions on Japanese space development will be made based on this statement. In the Basic Plan on Space Policy, Japan lists three major objectives to be achieved in the future.

1. Ensuring space security
2. Promoting space utilization in the consumer field
3. Strengthening and maintaining the space industry infrastructure, science, and technology

A small item related to the 1st objective is ensuring the stable utilization of outer space. The Japanese government must also enhance the resilience of its space infrastructure.

The first step to enhance this resilience is to understand the whole system. In order to do this, we propose an ontology especially designed for a weather observation system. Visualizing the system will make it possible to clearly see the whole picture of the relationships between individual systems such as a satellite and ground station. In the future, this ontology will tell us which part of the system lacks the appropriate resilience. For example, based on an analysis, a satellite manufacturing company will be able to design a satellite with enhanced resilience.

1.2   Objective

The objective of this paper is to develop an ontology to understand the entire space infrastructure, especially the weather observation system, and to acquire insights about what to do next in the space infrastructure development. The development of this ontology will make it possible to determine whether the system has resilience or not. Based on the results, we can obtain information about the steps needed to enhance the resilience of the system capabilities.

1.3   Related Works

James Martin proposed the conceptual framework for the entire Earth Observation System of The National Oceanic and Atmospheric Administration (NOAA) in the United States [6]. Figure 1 shows a diagram of all of the entity relationships in NOAA's Observing System Architecture (NOSA).

This diagram makes it easier to understand why it was previously difficult to understand what they had and what they were doing with what they had. This diagram allows the viewer to understand the whole system and the relationships with other system entities.

L.C. van Ruijven tried to develop a system engineering ontology to enable model-based systems engineering (MBSE) by creating a set of information models as defined in ISO 15926-11 [7]. This ontology makes it possible to solve the lack of

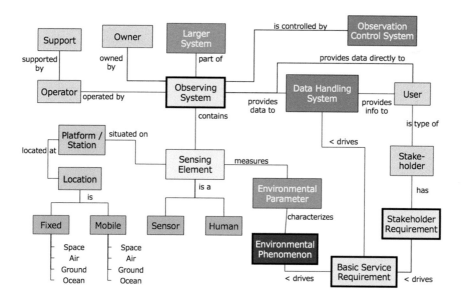

Fig. 1 Entity relationship diagram for NOAA's observing system architecture

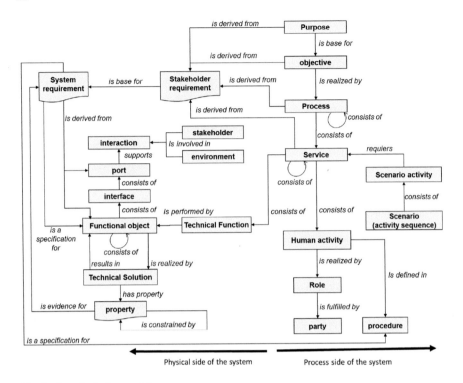

Fig. 2 Generic design model

interoperability and verbal chaos between the various parties involved in a project. He focused on the processes described in ISO 15288, including the stakeholder requirement definition process, requirement analysis process, and operational and maintenance process. He integrated the generic design model shown in Fig. 2 by developing an ontology that included each process.

As previously shown, the ontology approach is a very efficient method for determining the whole picture of each field at a highly abstract level. The ontology can be easily understood by people who are not familiar with the field. Thus, the ontology would promote understanding and beneficial discussions at the same abstraction level without any misunderstandings.

2   Research Process

In this research, we first developed an ontology for a weather observation system utilizing the Department of Defense Architecture Framework 2.0 (DODAF 2.0) and an enabler relationship. Then, we applied this ontology to a weather observation system and confirmed its resilience from the results.

In the future, by refining and using this ontology, we can develop a resilient system that can maintain the system capabilities.

3 Ontology

3.1 What Is an Ontology?

What is an ontology? An ontology identifies and defines concepts and terms [8]. An ontology provides descriptions of concepts and their relationships in a domain of interest. Furthermore, an ontology provides usage definitions, more than a dictionary or a taxonomy [9], which means an ontology is not just a classification system.

An ontology has two advantages. First, the user can obtain a consensus. When numerous parties are involved in a project, it is very difficult to obtain a consensus because there are many different interpretations of concepts. If they have the same dictionary containing definitions at a very deep conceptual level, misunderstandings between the parties can be prevented compared to the case when such a dictionary is lacking. Having a common dictionary makes it possible to explicitly express tacit knowledge. The second advantage is the ability to reuse and share knowledge. It is possible to identify the basic concepts that constitute such knowledge by considering the original object to be an object entity. Then, by considering the hierarchy according to the abstraction level of the knowledge, the user can consider the origins of this knowledge from the basics and find shared and reusable knowledge. Mainly, an ontology contains three types of relationships, as shown in Fig. 3.

(A) Class relationship between concepts
There are several categories of satellites, including earth observation satellites and communication satellites. Earth observation satellites also have sub-categories like weather satellites and intelligence satellites. The Himawari

Fig. 3 Three basic relationships of ontology

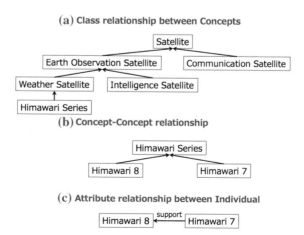

series refers to a kind of weather satellite. A Himawari satellite is a geostationary satellite, operated by the Japan Meteorological Agency (JMA), and supports weather forecasting, tropical cyclone tracking, and meteorology research.

(B) Concept-concept relationship
    The Himawari series includes eight satellites (Himawari 1–Himawari 8). Here is shown that Himawari 7 and Himawari 8 are part of Himawari series.

(C) Attribute relationship between individual satellites
    Currently, Himawari 8 is in operation, and Himawari 7 supports Himawari 8 as its backup. In the ontology description, we can write that "Himawari 7 supports Himawari 8."

3.2 How to Develop Ontology

Before introducing the weather observation ontology, we will discuss its framework. When we attempted to develop an ontology and capture the entire space system, we applied an existing framework (DODAF 2.0) [10]. DODAF is used by United States engineering and acquisition communities to describe the overall structure for designing, developing, and implementing systems [11]. DODAF provides a visualization infrastructure for the concerns of specific stakeholders organized by viewpoint. DODAF 2.0 utilizes eight viewpoints: the all viewpoint, capability viewpoint, data and information viewpoint, operational viewpoint, project viewpoint, services viewpoint, standards viewpoint, and systems viewpoint. DOD utilized this architectural framework to develop a perfect defense system without any omissions.

We applied this architectural framework to capture the entire weather observation system. We selected four viewpoints: the capability viewpoint, operational viewpoint, service viewpoint, and system viewpoint. These viewpoints have a relationship called an "enabler relationship." This enabler relationship is one of the relationships between viewpoints [12]. In Fig. 4, viewpoint 1 is enabled by viewpoint 2, and viewpoint 1 is utilized by viewpoint 2. This relationship can be applied to the DODAF 2.0 viewpoint. The capability is enabled by the service, and the service is enabled by the operation and system. Applying the enabler relationship to DODAF makes it possible to explicitly understand the layer structure.

In addition to the four-viewpoint structure, we decompose the system viewpoint in detail. In Fig. 5, we can clearly separate the roles of the function and physical viewpoints. This idea is similar to the systems engineering process in IEEE 1220 [13]. After the requirement analysis, we consider the functional analysis process and synthesis process. This decomposition makes clear the difference inside the system. It is also possible to apply the enabler relationship to the function viewpoint and physical viewpoint. We can clearly visualize this relationship.

By integrating Figs. 4 and 5, we can obtain the ontology for the weather observation system shown in Fig. 6.

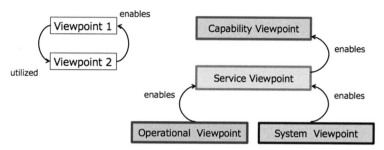

Fig. 4 Enabler relationship and its application to DODAF 2.0

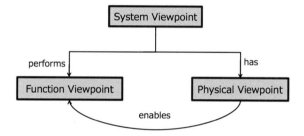

Fig. 5 Structure of system viewpoint and relationship

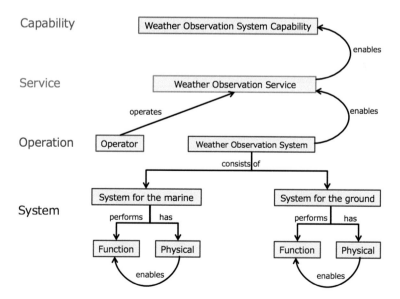

Fig. 6 Ontology for weather observation system

This figure shows the basic relationships between the viewpoints in this paper. In the system viewpoint, the weather observation system consists of the ground system and the marine system. Each system has a physical structure and performs functions. Of course, as previously mentioned, a physical structure enables functions. The weather observation system enables a weather observation service. The service is operated by an operator and enables the weather observation system capability. This is the whole picture of the weather observation system.

4 Applying Ontology to Weather Observation System

4.1 Result

Let us now apply the previously propose ontology to a weather observation system, particularly the Japanese weather observation system. The result is shown in Fig. 7.

We assume the situation that the weather observation system tries to forecast a typhoon. The weather observation system's ability to forecast a typhoon can be divided into forecasting the typhoon's track and forecasting its intensity. The main capability is enabled by the typhoon forecast service, which consists of providing a

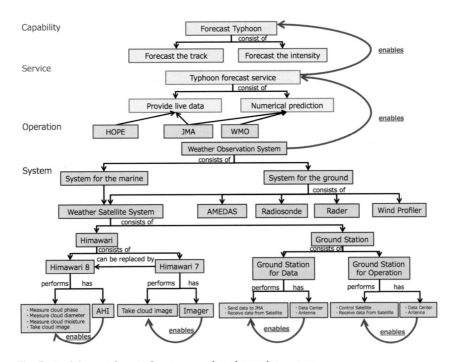

Fig. 7 Applying ontology to Japanese weather observation system

live data service and conducting numerical prediction. These services are operated by the Himawari OPeration Enterprise corporation (HOPE), JMA, and World Meteorological Organization (WMO). Then, the weather observation system enables the service. As shown in Fig. 7, the marine system is the weather satellite system, and the systems for the ground are the Automated Meteorological Data Acquisition System (AMEDAS), radiosonde, radar, and wind profiler on the weather satellite system consists of the Himawari satellites and ground station. The Himawari satellites consist of Himawari 8 and Himawari 7, and Himawari 8 can be replaced by Himawari 7. Thus, we define their relationship as "can be replaced by." There is a large performance difference between these two satellites. Himawari 8 has an imager called the Advanced Himawari Imager (AHI), which has more functions than Himawari 7. These functions include measuring the phase, diameter, and moisture of clouds, as well as acquiring images of them. These functions are enabled by physical structures. The ground station consists of two entities. The first is a ground station for data. It has a data center and antenna as physical systems. These make it possible to send data to JMA and receive data from the satellites. The second is a ground station for operation. It has the same physical systems as the ground station for data, but it has different functions for satellite control.

4.2 Considerations

As a result of applying the ontology to the Himawari system, we can show the entire Himawari system from the capability viewpoint, operational viewpoint, service viewpoint, and system viewpoint, which has both the function viewpoint and physical viewpoint. The model makes it easy to understand and obtain insights with the need to classify the entities and their relationships with the other entities. For example, Himawari is composed of two satellites, and Himawari 7 exists as a backup satellite for Himawari 8. If Himawari 8 suddenly stops working, Himawari 7 can be substituted to obtain cloud images. However, the other three functions (the cloud phase, diameter, and moisture measurement functions) cannot be carried out by Himawari 7. Thus, action should be taken to compensate for this lack.

In addition, we can consider the system resilience from Fig. 8. The definition of resilience was already discussed in Sect. 1.1. If the system of interest loses part of its architecture, can it maintain its capability? That is the question that we really want to answer.

Case 1: Lose a1
The system retains A and a2. Thus, the system performance becomes poor, and the redundancy becomes bad.
Case 2: Lose a2
The system retains A and a1. Thus, the system redundancy becomes bad.
Case 3: Lose A
Viewpoint X also loses a1 and a2. Thus, the entire viewpoint (X) will be lost.

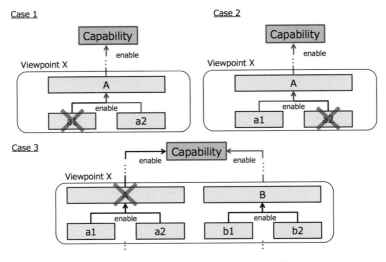

Fig. 8 Resilience in each case to maintain capability

In these three cases, case 1 and case 2 retain system capability. Of course, the performance and redundancy become bad in both cases, but the only concern is whether or not we can maintain the system capability. Hence, case 3 is the only catastrophic case that the system developer must avoid. If the system is in case 3, another entity (like entity B) needs to be developed to retain the system capability.

For example, we can conclude that a weather satellite plays a very important role for weather observation from Fig. 7. Assume a situation where this satellite is destroyed. Of course, the system capability will still be realized because this is case 3 (A: marine system, B: ground system), but the accuracy of the forecasting will become poor. This system ontology can provide many insights about system resilience.

5   Conclusion

We developed a weather observation ontology that covers the entire system utilizing DODAF 2.0 and the enabler relationship. This ontology shows the system in a structured way because every entity belongs to the capability, operation, service, or system category. By visualizing the system entities and their relationships, users can develop a common understanding of the system. Because the ontology is described in common words and simple relationships, it is easy for many people to understand, even those who are unfamiliar with weather observation systems. This weather observation ontology helps to bridge any gaps in knowledge or conceptual understanding. The developed ontology was verified to be of benefit in real systems by applying and using it to identify the vulnerabilities of a weather observation system.

The ontology also provided information about system resilience based on three cases. The model can assist with understanding the current system situation from the resilience perspective and in taking action to maintain the system capability when something happens to system entities.

6 Future Work

In future work, the ontology should be refined to clarify the intensity. The ontology could also be used obtain information to enhance the system resilience in both a qualitative and quantitative way. In order to realize this research, we should define how to measure resilience. This means we should quantify the resilience by setting a criterion.

References

1. Su, S.-Y., Kessler, D.J.: Contribution of explosion and future collision fragments to the orbital debris environment. Adv. Space Res. 5(2), 25–34 (1985) (ISSN 0273-1177)
2. Department of defense, fact sheet: DoD strategy for deterrence in space
3. Strategic headquarters for space policy, government of Japan: basic plan on space policy. (2013)
4. Department of defense, fact sheet: resilience of space capabilities
5. Pawlikowski, E., Loverro, D., Cristler, T.: Space: disruptive challenges, new opportunities, and new strategies. Strateg. Stud. Quart. 6(1), 45 (2012)
6. Martin, J.N.: On the use of knowledge modeling tools and techniques to characterize the noaa observing system architecture. In: INCOSE Proceedings, INCOSE Symposium (2003)
7. van Ruijven, L.C.: Ontology for systems engineering as a base for MBSE. In: INCOSE IW (2015)
8. Holt, J., Perry, S.: SysML for Systems Engineering 2nd Edition: a Model-Based Approach. The Institution of Engineering and Technology (2013)
9. Fosse, E.: Model-based systems engineering (MBSE) 101. In: INCOSE (2014)
10. Department of defense: DOD architecture framework version 2.02
11. Handley, H.A.H.: Incorporating the NATO human view in the DODAF 2.0 meta model
12. Shirasaka, S.: Architecture framework development process using meta-thinking and enabler. Keio University (2011)
13. ISO/IEC standard for systems engineering-application and management of the systems engineering process. ISO/IEC 26702 IEEE Std 1220-2005 first edition

Renewable Integration in Island Electricity Systems—A System Dynamics Assessment

George Matthew, William J. Nuttall, Ben Mestel, Laurence Dooley
and Paulo Manuel Ferrão

Abstract Island electricity systems tend to rely heavily on the use of fossil fuels for the everyday supply of customer needs, so there are both significant economic and environmental benefits from the decarbonisation of these systems. One such key global environmental benefit is the anticipated reduction in CO_2 emissions and its associated effects on climate change. In recognition, many islands are already pursuing ambitious goals for renewable energy sources integration. The resulting effects of policy on the long-term investment decisions however, need to be better understood. This paper presents a system dynamics simulation model which evaluates the adoption and diffusion of renewable generation sources within an existing island electricity system. In particular, renewable sources within the Azorean island of São Miguel are considered, with findings revealing that the requisite long-term investments are framed by the local experience of the renewable technologies and the pursuit of further renewable integration policy targets.

Keywords Renewable integration · Island electricity systems · System dynamics

G. Matthew (✉) · W.J. Nuttall · B. Mestel · L. Dooley
Faculty of Mathematics, Computing and Technology, The Open University,
Walton Hall, MK MK7 6AA, UK
e-mail: george.matthew@open.ac.uk

W.J. Nuttall
e-mail: william.nuttall@open.ac.uk

B. Mestel
e-mail: ben.mestel@open.ac.uk

L. Dooley
e-mail: laurence.dooley@open.ac.uk

P.M. Ferrão
IN+, Instituto Superior Tecnico, Av.Rovisco Pais, 1049-001 Lisbon, Portugal
e-mail: ferrao@tecnico.ulisboa.pt

© Springer International Publishing Switzerland 2016
M.-A. Cardin et al. (eds.), *Complex Systems Design & Management Asia*,
Advances in Intelligent Systems and Computing 426,
DOI 10.1007/978-3-319-29643-2_8

1 Introduction

This paper focuses upon the opportunities and challenges facing those making investment decisions for the integration of renewable sources; to obtain a better understanding of the resulting future portfolio of electricity generation mixes and the possible benefit of these to stakeholders. To streamline the research objective, attention is focused upon an island system where technical issues are small in scope and largely local, but the political economy is largely external. In [1], an extensive review of renewable integration into island systems is presented, which recommends that future models of such systems should incorporate both regulatory environments and the dynamics of learning curves of renewable sources, in order to comprehensively evaluate investment implications in the short, medium and long term. It is also important to identify the drivers and necessary investment and policy insights for a low-carbon optimized system, to achieve a sustainable future. Furthermore key stakeholders can elicit what are the distinct policy drivers and determine beneficial solutions and/or long-term investment strategies.

Several island-based studies have previously been conducted to investigate renewable integration challenges. For example, the island of Flores in the Azores has been used in the study in [2], as a "green" island paradigm characterized by high renewable energy penetration. A TIMES MARKAL analysis model was developed with exogenous demand growth, and a scenario-based approach used to find optimal solutions for energy system design and management, given the different possible exogenous 'evolutions' of electricity demand. The study analyzed the impact of demand side management (DSM) options, such as energy efficiency measures and dynamic demand response, showing that load shifting strategies could delay new investments, while rendering the current investments on renewable resources more economically viable. Of similar scope is the energy storage study in [3], again for the Azores, which used a least-cost unit commitment model analysis to determine the expected cost savings from introducing energy storage into existing electrical power grid networks. The study highlighted some challenges and revealed potential cost-savings from incorporating energy storage within a smart electrical power grid system. Another study [4], examined the use of electric vehicles for CO_2 emissions reduction by using renewable energy sources as the sole generation supply for charging the vehicles, with a least-cost economic dispatch and unit commitment model being proposed. Parness, also undertook a least-cost economic dispatch and unit commitment model. Recently [5–7] used system dynamics to analyze the distributed integration of renewable energy sources, carbon policy incentives and taxation within large interconnected energy networks. These works provide recommendations to policy makers for the uptake of, and pricing patterns for, tradable green certificates and carbon emissions prices. These models rely heavily on historical data, with the dynamics of the system being able to provide useful insights into these types of systems.

System dynamics is not an optimization methodology but rather it aids in understanding and gaining insights into complex systems, by capturing a system's key feedback structures and important sources of inertia and delays. Key endogeneities are often revealed which afford useful insights into the complex system structure and dynamic behaviors. System dynamics can elucidate scenarios and reveal hitherto unexpected behavior and phenomena in response to policies. The model presented in this paper highlights the necessary investment decisions needed to achieve the system's local renewable target; globally influenced CO_2 emissions targets; and to embed local learning experience of renewable technologies. Long-term sustainability policy interventions tied to these aspects of the system are also explored via the model. The model analyses the scenarios in which the rate of renewable integration is likely/not likely to be delayed, diluted, or defeated by unanticipated reactions and side effects. This becomes evident as the renewable goals are achieved and the learning curves reducing the cost of the renewable investments come into play.

Pruyt and Kwakkel [8] incorporated learning curves when using system dynamics to consider cost reductions accruing from the experience gained from previous installation of various competing energy technologies. The authors demonstrated the impact of learning curves on the cost of these competing technologies in energy transitions. In [9] the system dynamics approach was further applied to understand holistically the diffusion of a new technology, namely wind power. The authors showed the extent to which system dynamics captures the underlying mechanisms of diffusion processes and applied this to a large interconnected energy system. This provided the context for the case study into our chosen island: São Miguel in the Azores.

The rest of this paper is organized as follows: Sect. 2 presents the case study used for this work. The developed island renewable integration model is detailed in Sect. 3. Section 4 details the scenarios and Sect. 5 discusses the initial findings, analysis, validation and insights gained. The paper concludes with Sect. 6 and includes an outlook of the next steps and future modifications of the model.

2 Case Study: São Miguel—An Electrically Isolated Island

The Azores are an archipelago of nine Portuguese islands about 1,500 km west of mainland Portugal within the Atlantic Ocean. The islands are clustered into three major groups: the eastern, central and western groups, and they have a total population of 245,000 [10]. São Miguel is the largest island in the Azores, both in terms of size and population. It was chosen for this study because it is an electrically isolated island system that has ongoing extensive renewable technologies integration and plans for more in the face of high amounts of fossil fuel generation capacity, partly due to its carbon lock-in [3, 11].

São Miguel's electricity system is isolated in a technical, but not in a political and economic sense. The power system on the island is stand-alone without any interconnections to other islands or the mainland (preventing the import and export of electricity in peak supply and load situations). The island does not operate an energy market and it is dependent on the Portuguese mainland government to determine energy prices and policies [12, 13]. The local electricity company, *Electricidade dos Açores (EDA)*, serves all nine islands including São Miguel as a fully-regulated utility. São Miguel electricity customers pay the same retail electricity rates as mainland Portugal according to national law. Effectively, the Azorean electricity tariffs are subsidized by the rest of Portugal [3]. Policy requires that EDA follows least-cost planning procedures when investing in capacity additions or other grid enhancements. However, as highlighted in [14], and typical to most island electricity systems, São Miguel has a very large capacity reserve margin (well above 30 %). In contrast, the UK has a National Grid reserve margin goal of about 20 %. As a consequence of the high margin in São Miguel a significant amount of generation capacity is idle most of the year [15]. The annual electricity consumption load grew more than 3 % a year in the period 2005–2009, before dropping back after the global financial crisis (not featured in our model). Our model currently assumes that the future demand keeps rising by the same 3 % margin, with a fixed set of projected investments which will be responsible for covering additional electricity consumption [12]. The tariff prices for electricity are also expected to rise [13]. Issues of demand will be advanced in future work. In this study demand is exogenous and smoothly increasing.

A key aspect of this study is the recently imposed Portuguese national decree to achieve 75 % renewable electricity on the island by 2018, with an intermediate goal of 50 % renewable by 2015 [3]. However, there are no clear insights into the long-term dynamics for hastily adopted renewable policies. Would it be delayed, diluted or frustrated due to the global pressures of CO_2 emissions reduction or affected by the local learning curve of the renewable technologies? This paper provides clear insights into this issue as it details the key socio-techno-economic aspects of the renewable integration problem for a typical island electricity system.

3 Renewable Integration Model

The integration of renewables in the island electricity system raises many uncertainties and different types of complexities and dynamics. As with all complex systems, the structure of the system affects its behavior. The system dynamics approach proposed in this paper makes it possible to represent the dynamics of the system in terms of the "feedback" processes, stock and flow structures, time delays and accumulations. These dynamics arise from the interactions within the networked feedbacks (loops) of the system. There are various causal relationships between key system variables which can be either positive (+ve)/self-reinforcing or negative (–ve)/self-correcting feedbacks. Accumulations are the individual stocks or measurable quantities of the system, i.e., the accumulated CO_2 emissions, accrued cost of new renewable capacity and the installed renewable capacity. These characterize the state of the island system and also are the sources of inertia and memory. The flows, such as the investment rate and net monthly CO_2 emissions, are directly linked to their respective stocks and reflect the rates at which these stocks increase or decrease. In the modelling, planned renewable investments and existing renewable capacity are stocks which are determined endogenously. Our model is a long-timescale investment model and is not a short-term grid balancing model. We adopt a one-month time-step. As such in this work we are largely insulated from short-term issues of weather and renewables intermittency.

Figure 1 shows the three main feedback loops influencing renewable integration within the system. The green loop (balancing effect of locally influenced renewable target) captures the causal relationship between the amount of renewable capacity installed and the shortfall of the amount needed to reach the local renewables target.

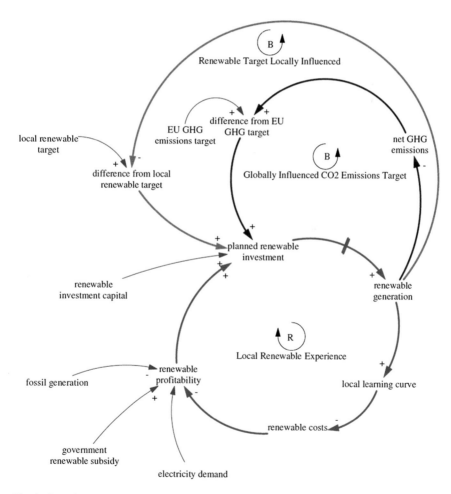

Fig. 1 Overview of the key feedback structures related to renewable integration within the island electricity system

The purple loop (balancing effect of the globally-influenced CO_2 emissions target) shows the effects of the installed renewable capacity and planned investments of renewables on the global emissions targets of the island system. The red loop (reinforcing effect of local renewable learning curve experience) captures the extent of cost reductions that accrue from the experience of installing renewables. This loop captures the breakeven cost of renewable production capacity that is required for the system to be sustainable. These three loops are the key components underpinning the model's structure, with their interactions being important for understanding the emerging characteristics of the system. In the work reported here we do not consider issues of long-term seasonal energy storage. This will be considered in future work focusing on the demand side. Shorter term (e.g. diurnal) storage is also omitted from this initial long-term investment model.

Additionally, social and economic impact plays a key role in the overall model, though at this stage these are secondary to the main feedback loops. The model crucially highlights a generic structure aggregating all renewable sources into a single entity. Later disaggregation will permit more accurate modelling, as the technological advancement with learning-curve cost reductions and profitability of the individual renewable sources can differ significantly. In later work we intend to focus on DSM and related social factors and to model this demand forecast as an endogenous component of the system. This will be pursued during further enhancements of the model.

The simulation model is implemented using the Vensim software package. The model has been derived from the causal loop diagram (Fig. 1), and includes the stock and flow variables that capture the key system structure. The important exogenous inputs are the local renewable targets, the CO_2 emissions targets, the electricity price and the electricity demand. In future versions of the model the electricity demand will be endogenous, but in this study our aim is to focus on renewable integration policies and cost reductions from installation experience. In this paper, the demand is assumed to be exogenous and storage is neglected.

To model cost reductions we follow [8] and write $C_{t+\Delta t} = C_t \left(\frac{X_{t+\Delta t}}{X_t} \right)^{-e}$, where C_t is the investment cost per MW at time t, X_t is the cumulative constructed capacity (including decommissions), and e is the learning curve parameter. The parameter $e = -\log_2 (p)$, where p is the progress ratio with $0 \leq p \leq 1$. A *progress ratio* of 90 % means that for each doubling of X_t there is a cost reduction of 10 %. Following [16, p. 338], the model was tested with several realistic progress ratios. For the relationship of the progress ratio to the economic concept of learning-by-doing please see [8 and 16, p. 338].

Within the model (Fig. 2) the key stocks are the planned renewable investments, the installed renewable capacity and the cost of new renewable capacity. The growth of planned investments in renewable capacity depends on the investment rate, which, in turn, is affected by the total capacity required to meet (i) forecasted demand load; (ii) the financial expectations of investors; and (iii) the CO_2 and local renewable targets. To model the influence of targets, we use the approach given in

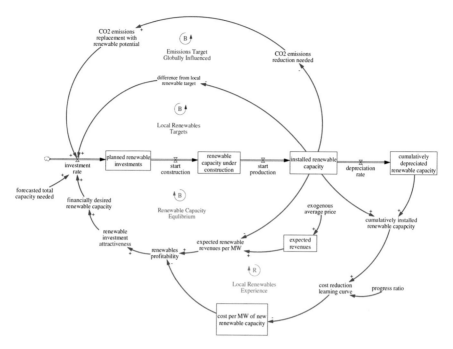

Fig. 2 Simplified stock and flow diagram of simulation model

[16], whereby R, the rate of adjustment of a variable S to a target S_*, is given by $R = (S_* - S)/T_A$, where T_A is the adjustment period.

Important dynamic components of the model are, for the monthly time step Δt:

a. the rate of change of installed renewable capacity, $\Delta C_I/\Delta t = P_C - P_D$, where P_C is the rate of commencement of generation and P_D is the depreciation rate;

b. the rate of change of planned renewable investments, $\Delta I_P/\Delta t = I - S_C$, where I is the investment rate and S_C is the rate of commencement of construction of new renewable capacity.

The investment rate I is a compound of several model variables:

$$I = \max((C_F - C_I)/T_I, (C_D - C_I)/T_I, R_R, R_E) + R_C$$

where C_F is the forecasted demand load; C_I is the installed renewable capacity; C_D is the financially desired renewable capacity; T_I is the capacity investment decision timeline; R_C is the rate of renewable capacity retirement; and R_R and R_E are, respectively, the rates of adjustment to the renewables and CO_2 emissions targets, as described above. The desired renewable capacity $C_D = A_R C_I$, where A_R is the investment attractiveness, which, following [17], we model as a piecewise linear function of profitability.

Structural validation of the model is achieved by comparing model outputs with historical output data of the real system for the endogenous installed renewable capacity. Historical data for 2005–2014 of the key exogenous variables, such as the demand load and electricity prices, have been used to determine appropriate data extrapolations using the Vensim SMOOTH and FORECAST functions.

To gain some insights into the next 35 years of the system structure and variables, the model has been implemented, with a monthly time step, for the period 2005–2049. The investment decisions for the renewable integration within the system have been observed and insights given based on different renewable targets and CO_2 emissions policies (CO_2 avoidance trading certificates and a price for CO_2 emissions have been ignored). It is important to keep in mind that this renewable model is still to be integrated with a fossil-fuel capacity model [18] and a future demand-side-management model and therefore model outputs and insights are provisional.

4 Three Scenarios

Within the scope of this model we are able to see the effects of the renewable target and CO_2 emissions policy on the planned and installed renewable capacity within the system. The additional effects on the cost of renewable investments due to the cost-reduction from installation experience is also highlighted in the model. In all scenarios, the initial planned renewable investments, installed renewable capacity and cost of renewable investments are obtained from the historical data of São Miguel, for the initial time of January 2005. Model calibration time is given from January 2005 to December 2014, whilst the simulation period runs from 2015 up to 2049. Three different scenarios are used for evaluation, which vary according to the desired policies. The extrapolated input data for the exogenous peak demand of the system and the initial cost of renewable investments remain the same in all scenarios. The CO_2 emissions and renewable target policies are implemented as stated in Sect. 3, and by fitting the adjustment time and required goal to the desired policy. The three scenarios are:

Reference scenario: This scenario considers the "business as usual" case and represents the most likely outcome under a midterm goal of 50 % reduced CO_2 emissions and 50 % installed renewable capacity targets within the system by 2030.
Less-aggressive renewable scenario: This scenario features renewable policies that have a goal of 50 % reduced CO_2 emissions and 50 % installed renewable capacity targets within the system, by 2050.
Aggressive renewable scenario: This scenario represents the goal of 50 % reduction in CO_2 emissions and 75 % installed renewable capacity within the system by 2018.

5 Analysis

Figure 3 shows the observed trend for the planned renewable investments within the system. In all three scenarios, as outlined in Sect. 4, the initial state of the model sets planned renewable investment at 9 MW reflecting the reality of São Miguel in 2005. The monthly planned renewable investments peaks somewhere after 2011 for all scenarios but as expected the policy of 75 % renewable by 2018 has a higher peak. After the peak the trend appears to be a steep decline into an exponential levelling off to zero around 2038 for all three scenarios. The similarity of the three scenarios is partly a consequence of assuming the same demand growth in each case. Future work will explore a range of scenario demands. Furthermore, it is expected renewable capacity will converge to meet the policy target as such, which is equivalent to archetypical s-shaped system dynamics behavior where such convergences might be with a system carrying capacity.

Following the planning stage, the installation and actual commisioning of installed renewable capacity can take 2–3 years. Figure 4 shows the amount of installed renewable capacity for all three scenarios, also compared to the real data of installed renewable capacity from 2005 to 2015 and the results of Ilić, in [19, Chap. 20]. All three of the scenarios reflect a similar amount of installed renewable capacity of about 39 MW for 2015 in line with the real data. The calibration time of the model was short and there was an initial deviation from the real data, however the long term trajectory of both our simulated model and the real data tends to be correlated. Note, Ilić achieved similar results to the 80 MW approximate value of installed capacity in 2028 using a stochastic dynamic programming method for long-term capacity planning. This provides some confidence in the validity of our work. The aggressive 2018 renewable policy has an installation peak that occurs faster and is higher than the 2030 50 % policy and the 2050 50 % policy. However

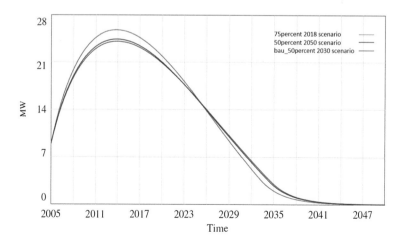

Fig. 3 Planned renewable capacity investments in 3 scenarios

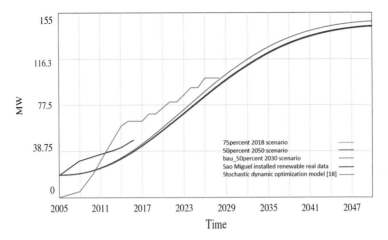

Fig. 4 Installed renewable capacity. Our modelling (3 Scenarios), real world data and independent modelling [18]

the final capacity in 2050 does not differ by much and we think that this can be attributed also to the electricity demand growth on the island (the carrying capacity of the system).

Figure 5 highlights the cost reduction learning curve. This study used a 90 % *progress ratio* resulting in a 10 % cost reduction on the initial cost price in 2005 for every doubling of the renewable capacity within the system. The new renewable cost price for 2005 as given by [20] was used. By 2050, the cost of new renewable capacity is shown to decrease by a small amount in all three scenarios indicating that the learning experience of the renewable element within the island is not very high. The corollary is that the learning-by-doing opportunity on such small islands

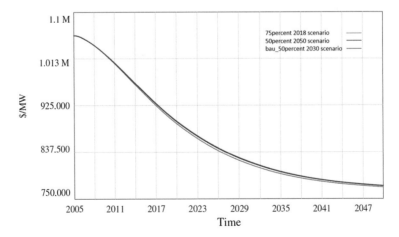

Fig. 5 Costs per MW of capacity in 3 scenarios illustrating the effect of learning-by-doing

is not very significant. Such learning effects are typically measured in terms of the cost reductions that can be expected from a doubling of installed capacity. The small size of the island system evidently restricts the potential for limitless growth and hence learning-by-doing.

Figures 6 and 7 show the renewable capacity needs for CO_2 emissions replacement and how the deviation from the local renewable target influences the three scenarios. In all cases, values are initially high then decrease in proportion to the aggressiveness of the associated policy. If stakeholders only considered these factors then there would be overly costly investment in the early years of the system. In Fig. 6, both the reference "business as usual" and less aggressive policies achieved the local renewable targets by 2023. However, we see indications that the aggressive 75 % 2018 policy appears to struggle. In that case the simulated model

Fig. 6 Convergence of modelled capacity with policy in 3 scenarios

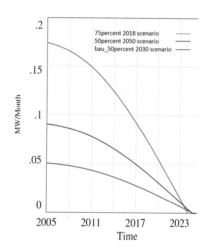

Fig. 7 Convergence with CO_2 emissions target in 3 scenarios

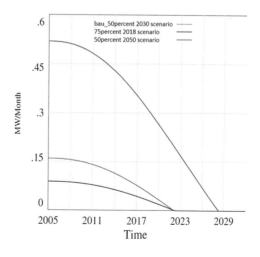

achieved its target by 2029. This can also be attributed to the carrying capacity of the system and the financial limitations attached to higher investments over a shorter time. Figure 7 implies that a higher amount of renewables are needed on a monthly basis for the aggressive renewable scenario in order to achieve the EU influenced CO_2 emissions goals. However, with less aggressive goals this target is achieved about 2 years later than the more aggressive policy goals. These observations emerge from this initial study restricted in scope and may evolve further as other factors are made endogenous to the model. One consideration that could greatly affect renewable generation is the effectiveness of diurnal and seasonal energy storage. We note the special role of hydro-power in this regard, which will be accounted for as different renewable types are disaggregated in future work.

6   Conclusions

This paper presents a system dynamics assessment of the renewable integration within the isolated island electricity system of São Miguel. Key components of the model highlight the cost reduction due to local learning from renewables and the type of renewable policy employed. Results and evaluations are starting to suggest that a sharp focus on achieving an aggressive renewable integration policy may lead to a boom and bust cycle of planning with periods of too much capacity. As shown in the results, in the long run, the required renewable targets will be achieved eventually, even with less aggressive renewable policies. This implies that possibly the financial health of the system can be jeopardized due to higher investment costs needed over a short period of time. Island systems typically suffer from a weaker innovation landscape and from limited opportunities for learning-by-doing. These realities combined with the preliminary results reported here suggest that island systems policy-makers should consider adopting a carefully-paced approach and should probably avoid establishing a world-leading position in innovation. That said, the small scale of island systems can lend themselves to experimentation and world-class opportunities from learning-by-research [21]. All these initial ideas will be re-evaluated in the light of the more holistic work to come.

The insights distilled from the model show that there are benefits to be obtained from considering all of the key feedbacks including installation experience and the urgency of the renewable capacity targets as renewables are integrated. These seem to steer the long-term renewable investment outlooks and provide the gaps for the optimal generation mixes of the system. The model of the system uses many key variables such as demand forecast and electricity tariffs as exogenous inputs. Future work will include analysis of the demand forecast within the model with a view to modelling more aspects endogenously. Storage and renewable disaggregation will have a particular role to play in future work. We intend to consider shorter time-steps and in this way to model the daily/weekly short-term drivers of renewable dispatch within such island electricity systems. This will be based on the short-term availability and the relative cost of renewables compared to other electricity options within the island system. We further intend to include energy efficiency/demand side response and diurnal storage in our modelling.

Acknowledgements We are most grateful to Pedro Carvalho, Stephen R. Connors, Manuel Heitor, Maria Ilić, João Martins de Carvalho, André Pina for helpful advice and discussions. The views expressed are those solely of the authors.

References

1. Weisser, D.: On the economics of electricity consumption in small island developing states: a role for renewable energy technologies? Energy Policy 32(1), 127–140 (2004)
2. Pina, A., Silva, C., Ferrão, P.: The impact of demand side management strategies in the penetration of renewable electricity. Energy 41(1), 128–137 (2012)
3. Cross-Call, D.F.: Matching energy storage to small Island electricity systems: a case study of the Azores. Master's Thesis MIT (2013)
4. Parness, M.: the environmental and cost impacts of vehicle electrification in the Azores. Master's Thesis MIT (2007)
5. Ford, A., Vogstad, K., Flynn, H.: Simulating price patterns for tradable green certificates to promote electricity generation from wind. Energy Policy 35(1), 91–111 (2007)
6. Ford, A.: Simulation scenarios for rapid reduction in carbon dioxide emissions in the western electricity system. Energy Policy 36(1), 443–455 (2008)
7. Ford, A.: Greening the economy with new markets: system dynamics simulations of energy and environmental markets. Int. Conf. Syst. Dyn. Soc. 1–26 (2010)
8. Pruyt, E. Kwakkel, J.: Energy Transitions towards Sustainability I: A Staged exploration of complexity and deep uncertainty. In: Proceedings 29th International Conference of the System Dynamics Society, pp. 1–26 (2011)
9. Bildik, Y., Van Daalen, C.E., Yü, G., Ortt, J.R., Thissen, W.A.H.: Modelling wind turbine diffusion: a comparative study of California and Denmark 1980–1995. In: International Conference of the System Dynamics Society, pp. 1–25 (2015)
10. Azores: Azores (2015). http://www.azores.com/. Accessed 18 Mar 2014
11. MIT-Portugal: MIT-Portugal (2013). http://www.mitportugal.org/research-overview/research.html#modelsdesign. Accessed 06 July 2014
12. EDA: Caracterização da Procura e da Oferta de Energia Eléctrica 2009–2013 (2008)
13. ERSE: Plano de Promoção da Eficiência no Consumo de Energia Eléctrica para 2012–2013 (2012). http://www.erse.pt/eng/electricity/tariffs/Paginas/default.aspx. Accessed 18 Mar 2014
14. Perez, Y., Ramos Real, F.J.: How to make a European integrated market in small and isolated electricity systems? The case of the Canary Islands. Energy Policy 36(11), 4159–4167 (2008)
15. EDA: EDA (2014) http://www.eda.pt/Paginas/default.aspx. Accessed 18 Mar 2014
16. Sterman, J.D.: Business Dynamics: Systems Thinking and Modeling for a Complex World (2000)
17. Black, J.: Integrating demand into the U.S. electric power system: technical, economic, and regulatory frameworks for responsive load. PhD Thesis MIT (2005)
18. Matthew, G.J., Nuttall, W., Mestel, B., Dooley, L.: Insights into the thermal generation futures of isolated island electricity systems using system dynamics. In: International Conference of the System Dynamics Society, pp. 1–17 (2015)
19. Ilić, M., Xie, L., Liu, Q.: Engineering IT-enabled sustainable electricity services: the tale of two low-cost green Azores Islands. Springer Science and Business (2013)
20. IEA: Projected Costs of Generating Electricity, vol. 118, Suppl no. 7 (2010)
21. Jamasb, T., Nuttall, W.J., Pollitt, M.: The case for a new energy research, development and promotion policy for the UK. Energy Policy 36(12), 4610–4614 (2008)

A Systematic Approach to Safe Coordination of Dynamic Participants in Real-Time Distributed Systems

Mong Leng Sin

Abstract Computer systems employing autonomous robots have been demonstrated in many areas including defense and homeland security. Some of these systems are safety critical; there are safety requirements linking to human safety or crucial infrastructure. Such safety critical systems must be reliable despite having unreliable components (e.g., communication, sensors, or actuators). In addition, mobile robots may move in and out of the operational area (dynamic participant), resulting in the complexity of coordination amongst varying robot numbers. This paper presents Comheolaíocht—a systematic approach to design coordination protocols for dynamic participants in real-time distributed systems allowing autonomous mobile robots to achieve their goals safely. The three-step approach begins with system modeling to capture the system specification. The second step, system analysis, analyses the system specifications to determine whether Comheolaíocht can provide a reliable solution and if possible, outlines a coordination strategy. The third step, protocol derivation, derives the system's coordination protocols which guarantee robots' exclusive access to resources despite imperfect communication. The process is applied to a practical application in which autonomous vehicles coordinate to drive safely on a road.

1 Introduction

Advances in robotics and mobile communication have enabled the development of autonomous mobile robot systems. Instead of using a single complex robot, the potential advantages of deploying multiple autonomous mobile robots are numerous

M.L. Sin (✉)
DSO National Laboratories, 20 Science Park Drive, Singapore 118230, Singapore
e-mail: smonglen@dso.org.sg

© Springer International Publishing Switzerland 2016
M.-A. Cardin et al. (eds.), *Complex Systems Design & Management Asia*,
Advances in Intelligent Systems and Computing 426,
DOI 10.1007/978-3-319-29643-2_9

121

[1, 2]. A multi-entity system's intrinsic parallelism provides robustness to single entity failures, and in many cases, can guarantee better time efficiency [1]. In addition, some tasks may be inherently too complex or impossible for a single robot to accomplish [2].

Computer systems that employ multi-entity coordination have been demonstrated in many applications including intelligent transportation systems [3–6], robotic soccer [7, 8], and construction [9, 10]. On one hand, each of these applications must achieve their own functional requirements such as scoring goals in robotic soccer or completing a construction. On the other hand, non-functional requirements like reliability are just as important [11].

Most multi-entity systems have their own specification of reliability. For instance, autonomous underwater vehicles in a mine-countermeasure mission must find all the mine-like-objects despite vehicles' failures [12, 13], and autonomous vehicles in an intelligent transportation system must not crash [4, 14]. Such multi-entity systems may also have some *goals* that must eventually be satisfied or some *safety constraints* that must be satisfied at all times. The autonomous underwater vehicles eventually discovering all mine-like objects is an example of a system's goal and vehicles not crashing is an example of a system's safety constraint. The combination of many different entities may result in a complex system where the entities' goals and safety constraints are interdependent. To ensure successful operation, these autonomous robots must coordinate their behaviors towards achieving the goals while respecting the safety requirements.

This paper presents a three step approach to design coordination protocols for mobile robots that may move in and out of the operational area (dynamic participant), in real-time distributed systems allowing autonomous mobile robots to achieve their goals safely. We fondly called this approach Comheolaíocht, a shorten name for comhordú modheolaíocht; Coordination Methodology in Irish. Comheolaíocht begins with system modelling (Sect. 3) to capture the system specification. The second step, system analysis (Sect. 4), analyses the system specifications to determine whether Comheolaíocht can provide a reliable solution and if possible, outlines a coordination strategy. The third step, protocol derivation (Sect. 4), uses our coordination protocol CwoRIS [15] which guarantees robots' exclusive access to resources despite imperfect communication. This paper demonstrates the use of Comheolaíocht using a practical application in which autonomous vehicles coordinate to drive safely on a road. Autonomous vehicles crossing an intersection are an example of a scenario with dynamic participants; the vehicles may move in and out of the road at high speed. The autonomous vehicles scenario has strict requirements on safety and real-time properties.

2 Related Work

2.1 Intersection Collision Avoidance

Autonomous vehicles crossing an intersection were first explored by Naumann et al. [16], who suggested passing a token among the vehicles to coordinate crossing. More recently, Dresner [4] proposed the managed intersection with a centralized intersection manager to take requests from incoming vehicles, calculates their trajectory and performs resource reservation. While centralized solutions are easier to optimize, they are vulnerable to single points of failure and require dedicated infrastructure, which might be expensive. There are also solutions that propose to group communication protocols which assume the number of continuous message loss is bounded [17, 18]. These assumptions are not realistic.

2.2 Real-Time Coordination Middleware

The main function of a middleware is to facilitate the communication and coordination of components distributed across several networked hosts, it should enable application engineers to abstract from the low-level details of network communication, coordination, reliability, scalability and heterogeneity [11]. Nine middleware systems were surveyed [19], and only two: LIME [20] and STEAM [21] support scalability and mobility. LIME [20], supports the mobile entities by having each entity maintains its own information. When an entity becomes connected to a group, a message is sent to all entities in the group to freeze their operations for information synchronization. LIME's assumptions that the network is not highly dynamic, can sustain a connection during a transaction, and having to freeze every entity whenever there is an entry or an exit is not practical in an environment with dynamic participants.

The real-time version of STEAM [21] and is implemented as space elastic adaptive routing, SEAR [22, 23]. SEAR provides real-time routing and feedback based on the TBMAC protocol [24]. Comhordú [14] uses SEAR and TBMAC to provide real-time coordination of mobile entities. Comhordú's coordination protocol ensures system reliability by allocating responsibilities to entities; such that for each specified safety-constraint, a particular entity is made responsible not to violate the constraint. The responsible entity maintains safety either by changing its behavior, delaying its action or requesting other entities to change their behavior. While the Comhordú-SEAR combination allows dynamic participants, support for it is limited by the possible break in communication whenever a node moves across the cell-based messaging protocol that SEAR is built on.

Similar to Comhordú, this work also depends on the underlying communication protocol; it uses the communication protocol Vertigo [25] which is able to supply the identities of entities within a specific area at a fixed instant.

3   System Modeling and Specification

System modeling is the first of Comheolaíocht's three steps in designing a solution
for a coordination problem. This stage involves step-by-step modeling and specification of the system by describing the different participants and their behaviors,
the environment, and the constraints between them.

Figure 1 shows the intersection crossing environment this paper uses as an
example for developing coordination protocols for autonomous vehicles to cross.
Furthermore, the junction scenario is modelled as a grid of *shared resources*;
vehicles crossing the junction must reserve their required resources using the
protocol in Sect. 5. Due to space limitation, this paper only specifies the most
important parameters. Interested reader can refer to the full step in [27].

4   System Analysis

The second step, system analysis, analyses the specifications captured in the first
step to obtain two results: firstly, it determines whether Comheolaíocht can provide
a reliable solution to a multi-entity coordination system, and secondly, if a reliable
solution exists, the step provides a coordination strategy that ensures the safety in
the system.

4.1   Modes

An entity's *mode* is a disjoint subset of the entity's states. An entity's *state* defines
the situation of an entity at a given time; the entity's state can be described by the

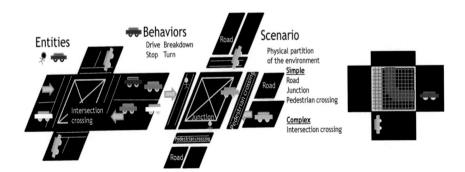

Fig. 1 Intersection crossing environment with its entities and behaviors

values of its set of variables. For instance, a developer may define the modes for an autonomous vehicle, with variables acceleration (a) and velocity (v) as:

Travelling : $a \geq 0, v \geq 0$ **Stopped** : $a \in range(a)$, $v = 0$ **Decelerating** : $a < 0, v$ $\in range(v)$

4.1.1 Fail-Safe Modes

A mode m of an entity x is said to be a *fail-safe mode* if and only if all of the states in m respect the safety constraints.

This definition of fail-safe mode reflects a snapshot of time and does not consider the implication of what may happen in the future. Consider the case where the developer defines a mode **MovingBeforeJunction**. In this mode a vehicle is approaching the junction at some positive speed. Since the safety constraint is only defined for vehicles inside the junction, an entity is not required to act safely in the **MovingBeforeJunction** mode and it is by definition a 'fail-safe' mode. However, the positive speed constraint in **MovingBeforeJunction** mode implies that the vehicle may only remain in this mode for some finite period before it enters the junction and possibly becoming unsafe. While it is sufficient for an entity to remain in a fail-safe mode so as to ensure that it do not violates the safety constraints, it may not be possible for the entity to remain in a fail-safe mode forever. Thus, a *long-lasting fail-safe mode* (LLFSM) is defined as: "A fail-safe mode is a long-lasting fail-safe mode (LLFSM) if the entity can remain in that mode for a long enough period such that the entity is guaranteed successful coordination."

In the definition, successful coordination means that the entity has obtained exclusive access to the shared resources that it requires. Successful coordination is required to ensure that the entity does not violate the safety constraint when it leaves the LLFSM and enters a non-fail-safe mode. The duration required to achieve successful coordination is dependent on various parameters such as the coordination protocol, number of entities involved and the lower-level communication guarantees available. An alternative interpretation for 'long enough' is that the entity must take some deliberate action to leave a LLFSM.

4.1.2 Mode Transitions

A *mode transition* from a mode x, to another mode, y, happens when an entity's state changes states from s_0 to s_1, such that $s_0 \in x$ and $s_1 \in y$.

Using the same autonomous vehicle example, a vehicle can transit from the **Stopped** mode to the **Travelling** mode by accelerating, changing a from 0 to positive. Mode transitions are categorized into three groups based on the entity's ability to control the cause, they are:

- **Controlled** mode transitions are caused by the entity's deliberate actions; that is, the entity can choose not to perform the actions. For example, a vehicle can decide to apply brakes to transition from the **Traveling** to **Stopped** mode.
- **Uncontrolled** mode transitions occur when some events happen that cause the transition without check. For example, a vehicle breakdown causes the entity to transit from **Traveling** to **Breakdown** mode.
- **Timed** mode transitions are a special subset of uncontrolled mode transitions that are caused by time passing. Mode transitions in this sub-group are special because although an entity may predict when the transition will occur (assuming that the entity has an internal clock), the entity cannot prevent the transition unless it can take some action to prevent it. For example, the mode transition from **TravelingBeforeJunction** to **TravelingInJunction** happens with the passing of time if the entity maintains its traveling action.

An entity's *mode transition diagram* describes all of the entity's modes and their possible mode transitions. Figure 2 shows an example of a vehicle entity's mode transition diagram for crossing an intersection. The identified modes are **TravelingBeforeJunction, StoppedBeforeJunction, FailedBrakes, InJunctionTraveling, AfterJunction** and **InJunctionBreakdown**. Each transition is named, and preceded by the transition causes of either **U, C** or **T**.

4.2 Coordination Strategy

Figure 2 shows an entity with failed brakes crossing an intersection. In the diagram, an uncontrollable transition from **TravelingBeforeJunction** leads into the **FailedBrakes** mode representing the situation of a brakes failure being discovered when the vehicle tries to decelerate. A timed transition from the **FailedBrakes** mode leads into the **InJunctionTraveling** non-fail-safe mode, showing that the vehicle is still traveling at some velocity and inevitably enters the junction. In this example, due to the possibility of a brakes failure the entity cannot deterministically

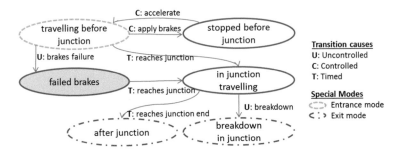

Fig. 2 Mode transition diagram of vehicle with failed brakes crossing an intersection

Table 1 Entity transition along an edge

Edge under examination	Other edges from the same vertex	Can entity transition along edge under examination deterministically?	Do the transition has deterministic transition time?
Uncontrolled	Some timed/uncontrolled	No	–
	All controlled	Yes	No (possibly infinite)
Timed	Some timed uncontrolled	No	–
	All controlled	Yes	Yes
Controlled	All controlled/timed	Yes	Yes
	Some uncontrolled	No	–

access its LLFSM despite having one; Comheolaíocht cannot ensure a safe solution for such systems.

In order for a mode transition diagram to be solvable in Comheolaíocht, an entity must be able to enter some LLFSM deterministically before its transition into a non-fail-safe mode.

Table 1 shows an analysis for entity transition with the third column specifying whether the entity can deterministically transition along the considered edge and the fourth column specifies whether the entity can calculate the transition time deterministically.

By using Table 1 and a suitable graph walk algorithm, Comheolaíocht can determine whether a particular scenario is solvable by examining each mode in every entity's mode transition diagram. Furthermore, the graph walk algorithm can be modified to output the actions an entity takes in order to be safe in a scenario. The details for such a graph walk algorithm can be found in [26].

4.2.1 Composition of Scenarios

The analysis for modes and mode transition diagrams focuses on a single scenario. In order to compose the separated scenarios into the application environment, the mode transition diagrams are evaluated to determine (1) the modes for which an entity transit into when it enters another scenario (e.g., from **road** to **pedestrian crossing** to **junction** and back to **road**), and (2) that the pre-conditions of a scenario can be satisfied by an entity entering some LLFSM of the previous scenario.

5 Protocol Derivation

The final step in Comheolaíocht's uses our CwoRIS pattern [16] to implement entities coordination. Using the environment model and constraints from step one, and the coordination strategy from step two, the CwoRIS pattern derives a protocol to provide distributed scheduling or mutual exclusion for dynamic participants that ensures exclusive access to resources despite imperfect communication.

5.1   CwoRIS Pattern

The CwoRIS pattern is described using its three-parts, (i) the lower-layer multicast protocol requirement, (ii) participants' *responsibility*, and (iii) the *request/feedback* protocol.

5.1.1   Protocol Requirement

The CwoRIS pattern is designed to operate on top of a multicast protocol that provides ordered delivery of messages, bounded message latency and real-time feedback. Ordered delivery of messages ensures that messages that are delivered to some nodes arrive at these nodes in the same order. Bounded message latency ensures that messages are either delivered within a certain period or discarded. Feedback on message delivery ensures that a sender receives result on the entities that a message is delivered. Real-time feedback ensures that feedback on message delivery is delivered to the sender within a bounded time. Ordered delivery is essential to CwoRIS because allocations are granted based on first-come/first-served (FC/FS); an unordered sequence may cause two entities with conflicting requests to believe that they have won the FC/FS race and access the shared resource at the same time, which violates the exclusive access property. Real-time feedback is required because CwoRIS supports a realtime system where entities must make decisions based on the outcome of whether a message is delivered within a fixed period. When applied to physical entities, CwoRIS further assumes that the lower-layer communication protocol provides geocast. Several protocols satisfy our requirements [23, 26]. In addition to communication requirements, CwoRIS assumes that physical entities have (i) a speed upper bound, and (ii) sufficient control to use only the resources it requires; in the autonomous vehicle junction crossing scenario, the resources are grids in the junction (see Fig. 1).

5.1.2   Participants' Responsibility

A node in the CwoRIS protocol does not enter the critical section unless it can be sure that every other node that may concurrently access the critical section has given way to it. Bouroche [14] termed such behavior as being "responsible". Each node is responsible for respecting the safety constraint (not entering the critical section), and a node can only enter the critical section by transferring its responsibility to other nodes that have given way to it. CwoRIS is an extension of the contract without feedback protocol, one of the contract protocols explored in [14].

5.1.3 Request/Feedback Protocol

Figure 3 shows the CwoRIS Request/Feedback protocol. In CwoRIS, a node transfers its responsibility by sending a request to everyone that may be interested in the node's intention to enter the critical section (i.e., the **sendRequest** function), thereby granting allocations on a FC/FS basis. Any node that receives such a proposal (i.e., the **recieveRequest** function) agrees to the proposal and defers to the sender if it has lower priority or it has not previously sent a message for the same resources. The node defers by not accessing the critical section at the same time as specified in the proposal message. When the request has been delivered to everyone of interest (i.e., the **receiveFeedback** function), the sender establishes that it is safe to relinquish responsibility and may access the critical section.

Figure 4 shows the example of a vehicle in the intersection collision avoidance scenario. There are four steps to guarantee an entity exclusive access to the shared resources. In the definitions, **MsgLatency** represents the latency within which a request must be delivered and feedback about message delivery returned to the sender. Figure 4a shows the vehicle entity mode transition diagram before entering the junction and Fig. 4b shows the time line of the mode transitions with respect to the four steps. The four steps in the coordination scheme are:

Lurking: During this step, an entity listens and builds the situation picture, before arriving at its decision point.

The *decision point* is defined to be the last time/space at which a node must start changing its behavior in order to avoid accessing the shared resources.

Resource request sending: The second step uses the CwoRIS request/feedback protocol to send a request to the set of entities that might access the same resources at the same time. The request message is sent at least **MsgLatency** before the entity's arrival at its decision point.

Request delivery and feedback receipt: The request should be delivered and the sender should receive feedback on the delivery results, at most **MsgLatency** after the resource request is sent.

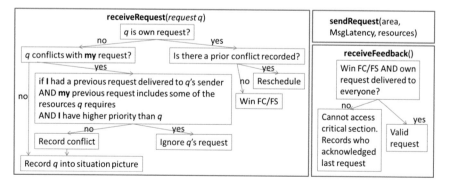

Fig. 3 CwoRIS request/feedback protocol

(a)

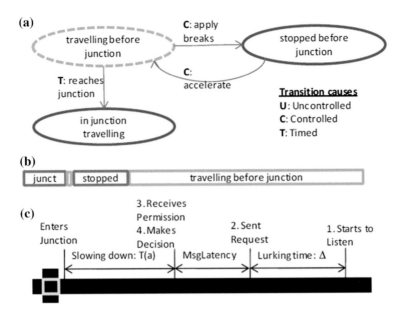

Transition causes
U: Uncontrolled
C: Controlled
T: Timed

(b)

(c)

Fig. 4 Steps for a vehicle in the intersection collision avoidance scenario

Decision making: Based on the result of request delivery, the entity may choose whether to access the shared resources or resend another request. There are two possibilities that can prevent an entity from acquiring access to its required resources: communication errors and the receipt of a conflicting request. Calculations of these parameters can be found in [15, 26].

6   Evaluation

The approach was evaluated against a complex intersection in the Dublin City (Fig. 5, left), and Fig. 5, right shows the composition of scenarios modelled in our evaluation.

Fig. 5 (*Left*) Intersection along Liffey River-Dublin (Picture from Google map). (*Right*) Liffey scenario composed by two simple intersection scenario

The simulation described in this Section is implemented using the Player/Stage. Player/Stage is used in the evaluation because of its support for multi-robots simulation and ease of usage. In the simulation, vehicles are randomly generated on a road leading into the junction. Each vehicle is equipped with a forward sensor providing the vehicle with the distance to obstacles in front of it. This sensor is used to for applying brakes when there is another in front of the vehicle. In addition, each vehicle is equipped with GPS and speed sensors.

Inter-entities communication is implemented such that messages sent are delivered to the entities based on the specifications (i.e., geo-cast with ordered delivery, bounded message latency and real-time feedback.) Imperfect communication is simulated by two parameters; the probability of a geo-cast failure and the probability of each message failing given that the geocast have failed. This second parameter is used to simulate delivery failure of a message or an acknowledgment.

The protocols are simulated and compared are:

- **Traffic lights** The traffic lights protocol models actual traffic lights in the intersection. The traffic light protocol is set to cycles of 20 s because it provides the best performances in the simulated scenario.
- **Centralized** The centralized protocol implements vehicles' reservation for crossing the junction using a centralized server modeled after [5].

Three other Comheolaíocht developed protocols are:

- **No optimizations** This protocol uses only local scheduling (without any optimizations), it implements the request/feedback protocol with inference and priority.
- **Delay resend** This version enhances the no optimization version by delaying an entity's resend request. A vehicle that has been waiting for a long time resends faster.
- **Preemption** This version implements the preemption-constraint.

The first comparison is a junction's throughput, which is defined as the number of vehicles going through the junction in an hour (Fig. 5). The Traffic lights protocol has the highest maximum throughput at 1100 vehicles per hour. The traffic lights protocol gives the highest throughput because the implementation naturally provides a 'platooning' effect where vehicles form platoons at the traffic lights which moves off at the same time. In contrast, all other protocols are based on reservation where vehicles are allocated higher inter-vehicle distances for reaction time in the event of vehicle breakdowns. In contrast, traffic lights protocol provides the least throughput when vehicle density is low; this is because a vehicle has to stop at the junction even when there are no vehicles crossing (Fig. 6).

Results for the simulation of communication errors and vehicle breakdowns are shown in Fig. 7. This simulation is performed only using the Comheolaíocht's Preemption protocol. As can be seen from the table, there are no collisions recorded in the simulations despite more than 14,000 breakdowns simulated with varying degrees of communication errors.

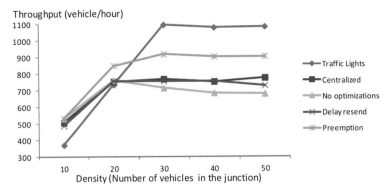

Fig. 6 Junction throughput with different implementations

Communication error	Vehicles	Average cross time	Maximum cross time	Breakdowns	Crashes
0%	19187	34.8	241.6	646	0
5%	18869	36.3	254.6	662	0
10%	18859	36.2	238	675	0
20%	18356	38.4	249	667	0
30%	18234	39.2	274.2	609	0
50%	17187	44.9	325	569	0

Fig. 7 Communication errors and vehicle breakdowns

7 Conclusion

Comheolaíocht provided systematic steps for developing protocols for reliable multi-entity systems. The process was applied to the intelligent transportation system and in particular the intersection crossing scenario.

Acknowledgment Special thanks are given to Prof. Vinny Cahill and Dr. Mélanie Bouroche, advisors of the author's PhD work in Trinity College Dublin, Ireland.

References

1. Dudek, G., Jenkin, M.R.M., Milios, E., Wilkes, D.: A taxonomy for multi-agent robotics. Auton. Robots **3**(4), 375–397 (1996)
2. Cao, Y.U., Fukunaga, A.S., Kahng, A.: Cooperative mobile robotics: antecedents and directions. Auton. Robots **4**(1), 7–27 (1997)

3. Dresner, K., Stone, P.: A multiagent approach to autonomous intersection management. J. Artif. Intell. Res. **31**(1), 591–656 (2008)
4. Dresner, K.: Autonomous Intersection Management. PhD thesis, The University of Texas at Austin (2009)
5. Dao, T.S., Ng, L., Clark, C.M., Huissoon, J.P.: Realtime experiments in markov-based lane position estimation using wireless ad-hoc network. In: Intelligent Vehicles Symposium, 2008 IEEE, pp. 901–906. IEEE (2008)
6. Qu, F., Wang, F.Y., Yang, L.: Intelligent transportation spaces: vehicles, traffic, communications, and beyond. IEEE Commun. Mag. **48**(11), 136–142 (2010)
7. Lau, N., Lopes, L.S., Corrente, G., Filipe, N.: Multi-robot team coordination through roles, positionings and coordinated procedures. In: IEEE/RSJ International Conference on Intelligent Robots and Systems, 2009. IROS 2009, pp. 5841–5848. IEEE (2009)
8. Mota, L., Reis, L.P., Lau, N.: Multi-robot coordination using setplays in the middle-size and simulation leagues. Mechatronics (2010)
9. Wawerla, J., Sukhatme, G.S., Mataric, M.J.: Collective construction with multiple robots. In IEEE/RSJ International Conference on Intelligent Robots and Systems, 2002, vol. 3, pp. 2696–2701. IEEE (2002)
10. Werfel, J., Nagpal, R.: Extended stigmergy in collective construction. IEEE Intell. Syst. 20–28 (2006)
11. Emmerich, W.: Software engineering and middleware: a roadmap. In: Proceedings of the Conference on the Future of Software Engineering, pp. 117–129. ACM (2000)
12. Sariel, S., Balch, T.: A distributed multi-robot cooperation framework for real time task achievement. Distrib. Auton. Robot. Syst. **7**, 187–196 (2006)
13. Sotzing, C.C., Lane, D.M.: Improving the coordination efficiency of limited-communication multi-autonomous underwater vehicle operations using a multiagent architecture. J. Field Robot. **27**(4), 412–429 (2010)
14. Bouroche, M.: Real-Time Coordination of Mobile Autonomous Entities. PhD thesis, University of Dublin, Trinity College (2007)
15. Sin, M.L., Bouroche, M., Cahill, V.: Scheduling of dynamic participants in real-time distributed systems. In: 2011 30th IEEE Symposium on Reliable Distributed Systems (SRDS), pp. 245–254. IEEE (2011)
16. Naumann, R., Rasche, R., Tacken, J.: Managing autonomous vehicles at intersections. IEEE Intell. Syst. Appl. **13**(3), 82–86 (2002). ISSN 1094-7167
17. Van Middlesworth, M., Dresner, K., Stone, P.: Replacing the stop sign: unmanaged intersection control for autonomous vehicles. In: Proceedings of 7th International Joint Conference on Autonomous Agents and Multiagent Systems, vol. 3, pp. 1413–1416 (2008)
18. Schemmer, S., Nett, E., Mock, M.: Reliable real-time cooperation of mobile autonomous systems. In: 20th IEEE Symposium on Reliable Distributed Systems, 2001. Proceedings, pp. 238–246. IEEE (2001)
19. Hadim, S., Al-Jaroodi, J., Mohamed, N.: Trends in middleware for mobile ad hoc networks. J. Commun. **1**(4), 11–21 (2006)
20. Murphy, A.L., Picco, G.P., Roman, G.C.: Lime: a middleware for physical and logical mobility. In: 21st International Conference on Distributed Computing Systems, 2001, pp. 524–533. IEEE (2001)
21. Meier, R., Cahill, V.: Steam: event-based middleware for wireless ad hoc networks. In: Proceedings of 22nd International Conference on Distributed Computing Systems Workshops, 2002, pp. 639–644. IEEE (2002). ISBN 0769515886
22. Hughes, B.: Hard real-time communication for mobile ad hoc networks. PhD thesis, University of Dublin, Trinity College (2006)
23. Hughes, B., Cahill, V.: Exploiting space/time trade-offs in real-time mobile ad hoc networks. Int. J. Mobile Netw. Des. Innov. **3**(1), 21–32 (2009). ISSN 1744-2869
24. Cunningham, R., Cahill, V.: Time bounded medium access control for ad hoc networks. In: Proceedings of the Second ACM International Workshop on Principles of Mobile Computing, pp. 1–8. ACM (2002)

25. Slot, M., Bouroche, M., Cahill, V.: Membership service specifications for safety-critical geocast in vehicular networks. In: Proceedings of 7th International Symposium on Communication Systems Networks and Digital Signal Processing (CSNDSP), pp. 422–426 (2010)
26. Sin, M.L: Comheolaíocht: a systematic approach to safe coordination of dynamic participants in real-time distributed systems. PhD thesis, University of Dublin, Trinity College (2013)

Foresight Study on Singapore Urban Mobility: Methodologies and Preliminary Insights

Seyed Mehdi Zahraei, Christopher Choo, Waqas Cheema
and Lynette Cheah

Abstract Singapore enjoys a world-class urban transportation system today, benefitting from a combination of long-term planning, continued investment in infrastructure, and readiness in adopting new technologies. However, planners will always need to prepare for uncertain futures, and understand driving forces and global trends that affect future urban mobility. In this paper, we present an ongoing foresight study on Singapore's urban transport and mobility up to 2030. Our objectives are to develop a shared understanding of the current state of the transportation system, highlight long-term challenges and opportunities, and establish networks between stakeholders. Through environmental scanning and expert interviews, our preliminary findings indicate that mobility-on-demand services, multi-modal transport, and e-commerce are the dominant future drivers of change in the urban mobility landscape. In addition, ageing population, growing population and travel demand, inefficiencies in urban freight, shortage in skilled manpower, and a general resistance to big policy changes are the key challenges facing Singapore urban mobility in future. Finally, in terms of technology, autonomous vehicles, real-time traveller information, and shared mobility are seen as potential game-changers for the future.

S.M. Zahraei (✉) · C. Choo · W. Cheema · L. Cheah
Engineering Systems and Design, Singapore University of Technology and Design,
8 Somapah Road, Singapore 487372, Singapore
e-mail: zahraei@sutd.edu.sg

C. Choo
e-mail: christopher_choo@mymail.sutd.edu.sg

W. Cheema
e-mail: waqas_cheema@sutd.edu.sg

L. Cheah
e-mail: lynette@sutd.edu.sg

© Springer International Publishing Switzerland 2016
M.-A. Cardin et al. (eds.), *Complex Systems Design & Management Asia*,
Advances in Intelligent Systems and Computing 426,
DOI 10.1007/978-3-319-29643-2_10

135

1   Introduction

This paper is a part of an ongoing foresight study on urban transport and mobility in Singapore. The project is in a series of foresight studies examining the Future of Cities at the Lee Kuan Yew Centre for Innovative Cities, Singapore University of Technology and Design.

Our core objectives are to develop a shared understanding of the current situation and issues in urban mobility in Singapore, facilitate future policy implementation by highlighting long term challenges and opportunities, and establish networks between experts and stakeholders in the mobility sector. Specifically, this study concerns only the land transport domain in Singapore, covering both passenger and freight sectors with government agencies and organisations in the transport domain as the audience.

The rest of the paper is organised as follows. First we develop the context by discussing the need for a foresight study on urban mobility in Singapore in Sect. 1. Next, we discuss our foresight methodology in Sect. 2. In Sect. 3, we discuss some of the preliminary insights we have gathered so far. We conclude in Sect. 4 by reviewing the insights and discussing our future work plan.

1.1   Singapore's Leading Role in Transport

Singapore has done well in developing a world-class urban transportation system over the past few decades, and plays a leading role in the region. In private transport, it enjoys the benefits of electronic road pricing, which was introduced in 1998 and helps to manage congestion by moderating the number of vehicles entering the central business district. In the area of autonomous vehicles, trials started in early 2015 to prepare the country for the future. As for public transport, the country's Mass Rapid Transit has made extensive use of driverless trains since 2003.

Despite the good quality of transport services in Singapore, uncertainties in the future cannot guarantee continued service quality. There is a need to pre-empt and plan for the future, meaning that planners should not only master daily operational issues, they must also develop a well-informed, long-term understanding for transport to address challenges that lie beyond the horizon.

1.2   Singapore's Context, Challenges, and Long-Term Planning

Singapore's cultural and political context differs substantially from many other high density cities in developed nations. The country's government is often in a strong position to pass legislation swiftly, develop the long-term vision for the country, and set in motion detailed development plans that are rapidly implemented. Thus, it is contingent on the government to ensure that its plans are realistic and practical.

A number of groups in the Singapore public service and academia are tasked with the responsibility to plan for the long-term future. Examples include the Centre for Liveable Cities at the Ministry for National Development, the Futures Division at the Ministry of Transport, and the Institute of Policy Studies, which is part of the Lee Kuan Yew School of Public Policy at the National University of Singapore. Other government agencies and research institutes also publish reports or conduct seminars to share thoughts about the future.

In the realm of transport, Singapore's Land Transport Authority was formed in 1995 to spearhead improvements to the land transport system, and published a white paper [9] to chart plans for the decade ahead. This was followed-up with the Land Transport Masterplan [10] that looked towards 2020, and the next revision [11], that planned for 2030.

These long-term plans for transportation were not drafted in isolation. The 1991 Revised Concept Plan by the Urban Redevelopment Authority plotted a 25-year development trajectory [21]. More recently, reports such as the 2013 Land Use Plan [14] and the Population White Paper [16] were released in succession, and each had core assumptions that were used in the 2013 Land Transport Masterplan. In this latest series of plans, Singapore's population is projected to grow to between 6.5 to 6.9 million by 2030, and massive developments are expected for Singapore's rail network, which will double in length to about 360 km by 2030.

Other studies have also covered transportation. The Economic Review Committee discussed advancing transportation infrastructure in strategic developments to catalyse growth in areas such as One-North, Tuas, and Jurong Island, and also provided a detailed set of recommendations on how to develop Singapore into a global integrated logistics hub [15]. More recently, to commemorate Singapore's jubilee celebrations, a vision for urban mobility was written by the Centre for Liveable Cities [19].

Over the years, the variety of studies and master plans has contributed immensely to the quality of transportation services present in Singapore today. However, there are two key challenges that policymakers must be mindful of when doing long-term planning.

First, the focus on policy solutions for single scenarios makes it difficult for planners to prepare for uncertainties. For example, the earlier Concept Plans based infrastructure provisions on the assumption of a fixed population target, which changed substantially with each revision. This had knock-on effects on the construction of transportation infrastructure, which takes many years to build. Policies need to be flexible to cater to uncertain futures, and responsive to swiftly address changing needs.

Second, many studies emphasise a prescriptive approach to policy, leaving the discussion about driving forces and global trends on the back-burner. For example, few would have anticipated that the rapid adoption of smartphones over the last few years would have led to the widespread use of mobile apps such as Uber, Google Maps, and GrabTaxi for transport. If car sharing and public transport become the dominant modes of transport over the next 15 years, fewer roads may be needed to handle traffic in the coming decades, and policymakers may decide to slow down road construction in anticipation of that future.

Are there opportunities for Singapore to take advantage of new technologies that could result in improvements of many orders of magnitude? Are current policies forward-looking such that they can deal with an uncertain future? Are there experts in the field who are working on potentially-disruptive technologies that policy-makers should start exploring today?

2   Foresight Methods for Singapore Urban Mobility

A foresight study focuses on developing multiple future scenarios to cover the spread of different possibilities that can occur as a result of today's decisions [7]. It assumes that the future is not an extrapolation of a set of predetermined trends and innovations. Instead, it builds on the principle of an uncertain future that can be shaped by today's actions. It is not the intention of foresight studies to predict the future correctly; but more to prepare stakeholders for a future that is inherently uncertain.

Typically, a foresight study uses a combination of different methods at different stages. Finding the right combination of these methods is one of the most important steps in the exercise. Some of the methods are quite simple to use and do not require significant expertise, while others are complex to develop and very time consuming. Furthermore, some methods are more practical when used for short-term projections, while others are better suited for long-term forecasts.

By evaluating more than 880 foresight studies conducted in Europe and other parts of the world, Popper [18] showed that on average five to six methods have been adopted for each foresight study. The process of selecting the most appropriate foresight methods for our study is a challenging task since there are more than 30 different foresight methods in literature and there is no specific guideline for sys-tematically selecting the methods for our foresight study.

Popper [18] classified foresight methods into four attributes based on their ability to gather or process information, namely *Creativity*, *Expertise*, *Interaction*, and *Evidence*. **Creativity** is about a mixture of original and imaginative thinking; methods relying on the inventiveness and ingenuity of individuals, or developed through brainstorming sessions. **Expertise** refers to the skills and knowledge of individuals in a particular area to make decisions, and provide advice or recom-mendations. **Interaction** recognises that expertise gains considerably from being brought together and challenged to articulate with other expertise. And finally, **Evidence** recognises that it is important to support analysis with reliable docu-mentation and measurement indicators.

After reviewing foresight methods in the literature, we decided to select our methods by considering the time horizon of the project and Singapore's context, and also by exploiting the four fundamental attributes of the foresight methods. The selected methods are as follows:

Environmental Scanning/Literature Review (Evidence attribute): Environmental scanning helps to understand the nature and pace of change in the environment, and to identify important economic, social, environmental, technological and political trends. Literature review represents a key part of the scanning process.

Expert Interviews (Expertise and Evidence attributes): Interviews are structured conversations which are used to gather insights from experts who are specialists in their respective fields.

Future Scenarios and Workshop (Creativity and Interaction attributes): Scenario planning is one of the most well-known and most cited technique for planning for the future. Edgar and Alänge [4] defined scenario planning as the process of creating several varied but plausible views (scenarios) of the future by considering the impact of uncertainties and driving forces. Scenarios help to identify future options and prepare stakeholders to tackle the world of uncertainties.

Technology Roadmapping (Expertise attribute): Technology roadmapping helps to identify critical technologies under development that can have game-changing effects on the system. The process of roadmapping helps to chart the course of these technologies and how they fit into future scenarios.

Apart from technology roadmapping, our selected methods have been frequently used in foresight studies as discussed in Popper [18]. However, we decided to include technology roadmapping since urban transport in Singapore will be significantly affected by technological developments in the future. It is worth mentioning that these are the main methods used in the transport and mobility foresight studies that we have reviewed so far, which we discuss later in this paper.

3 Preliminary Insights

3.1 Environmental Scanning and Literature Review

We first reviewed the Singapore land transport masterplans [9–11] and the Intelligent Transportation System (ITS) Smart Mobility Vision 2030. These documents helped us to understand Singapore's context and identify key areas that the government is focusing on for the future.

According to the land transport masterplan 2013, some of the key challenges Singapore faces include population growth that will increase mobility demand, more congestion on roads with a reduction in available space for new roads, and an ageing population. In addition, the ITS Smart Mobility Vision 2030 discussed areas of technological development that can manage demand and plan for seamless future mobility. These areas include real-time information availability, connected cars, shared mobility, enhanced traffic management systems for road pricing, autonomous vehicles, and green mobility.

Next, we reviewed foresight studies on urban transport/logistics. Singapore, as a city state, has its own set of land constraints. As an instance, residential areas coexist with business districts, parks and other commercial areas whilst in other megacities such as London, Seoul and New York, dense city centres are surrounded by suburbs. Nonetheless, many of these megacities have similar issues, such as increasing congestion and urban density, increasing and ageing populations, changing social behaviours, and dependency on combustion engines in automobiles. We also looked up other studies on general urban megacities in order to widen the scope of our literature review.

We reviewed twelve studies which were conducted by diverse groups, ranging from global consulting firms (e.g., Deloitte), independent international organisations (e.g., Forum of the Future), academic institutions (e.g., New York University), and government departments (e.g. New Zealand's Ministry of Transport). These studies cover time frames from 2018 to 2100, with a bulk of studies focused around 2030. The results are summarised in Table 1.

The results show that Future Scenarios, Expert Interviews and the Environmental Scanning/Literature Review are the most frequent methods used in the transport related foresight studies. Moreover, alternative energy/cost of energy, environmental sustainability, mobility-on-demand, virtual travel, vehicle automation, ageing population, and urbanisation are the dominant future drivers of change that urban cities around the world are focusing on.

3.2 Expert Interviews

Our study is ongoing, and we are conducting a series of interviews with experts from the government, academia, and industry professionals within the land transport sectors in Singapore. These interviews will help us to gather insight on current/future challenges, upcoming trends, industry evolution, and technology development.

Based on the interviews conducted so far, we identified a variety of challenges facing transportation in Singapore. These cover day-to-day operational issues such as managing peak hour traffic and resolving the first and last-mile problem, technological issues such as big data in intelligent transport systems, and global challenges such as climate change and urbanisation. We classified these challenges into four broad categories as follows:

Demographics, particularly in terms of the ageing population, was seen as a key challenge as senior citizens would need more assistance to get around. Also, the increasing resident population and foreign workforce will give rise to denser towns and the need to increase transport capacity.

Culture was also cited as a huge challenge, especially in terms of adopting more efficient modes of transport. The affluent population in Singapore considers cars as status symbols, shunning public transport and car sharing and cycling. Next, users have overly-high expectations of public transport in terms of cost and reliability. In

Table 1 Literature review on urban transport/logistics foresight studies

Foresight study	Foresight year	Cities/region	Foresight methods						Key drivers of the future
			Literature review	Expert interviews	Workshops	Future scenarios	Others[a]		
Townsend [20]	2030	United States Megacities	•			•			Cheap alternative energy, Vehicle automation, economic growth, individualism
van Voorst tot Voorst and Hoogerwerf [24]	2040	Dutch Urban Centers	•	•		•			Mobility-on-demand, virtual travel, legislation against free use of vehicles, vehicle automation
Intelligent Infrastructure Futures [17]	2056	United Kingdom Urban Centers	•			•	•		Environmental sustainability, public acceptance of intelligent infrastructure
DHL [2]	2050	Megacities		•	•	•	•		Energy price, environmental sustainability, political stability, global trade, individualism, robotics, rise of Asia, ownership to rental model
Martins et al. (2008)	2018	Brazil Urban Centers		•		•	•		Investment in infrastructure, efficiency of legislation and quality control
Auvinen et al. [1]	2100	Finland Urban Centers	•				•		Urbanisation, alternative energy sources, ITS integration, environmental sustainability
Zhao et al. [25]	2030	Jinan, China	•	•	•	•			Urbanisation, environmental sustainability

(continued)

Table 1 (continued)

Foresight study	Foresight year	Cities/region	Foresight methods					Key drivers of the future
			Literature review	Expert interviews	Workshops	Future scenarios	Others[a]	
Lyons et al. [12]	2042	New Zealand Urban Centers			•	•	•	Cost of energy, virtual travel
Gazibara et al. [6]	2040	Megacities		•		•		Resource scarcity, environmental stability, demographics, energy mix, governance model, values of future society
Zmud et al. [26]	2030	United States Megacities			•	•	•	Price of oil (energy), environmental sustainability, investment in infrastructure
Fishman [5]	2020	Megacities			•	•		Vehicle automation, virtual travel, ITS integration, mobility-on-demand
Ecola et al. [3]	2030	Chinese Megacities			•	•	•	Pace of economic growth, amount and type of constraints imposed on vehicle ownership and use, environmental conditions

[a]Technology Roadmapping, PESTE Analysis, System Dynamics, Consistency, Cross Impact, and Cluster analysis, and Delphi method

addition, environmental sustainability is typically ignored. As for telecommuting, the practice remains unpopular in Singapore even though it can take vehicles off the roads.

Regulation was discussed frequently over the course of our interviews. In the realm of autonomous vehicles, liability for accidents caused by vehicles of different levels of autonomy remains an issue. For public transportation, Singapore's design-build-operate-transfer model must align budgetary constraints with operational needs for maintenance and future expansion. For transport innovations such as Uber, regulatory frameworks still lag developments. And in terms of the next generation of electronic road pricing, authorities need to assure motorists about data privacy.

City Logistics or urban freight has been less studied, but is now a growing challenge. Inefficiencies in delivery arrangements, such as the congestion of freight vehicles at shopping malls, and re-deliveries for undelivered residential packages result in longer delivery times and the need for more drivers island-wide.

Expert interviews also helped us to identify several dominant trends that are likely to play key roles in how people commute in future. These include mobility-on-demand, multi-modal transport, e-commerce, the rising middle class, cycling, and teleworking, to name a few. In addition, we distilled technologies that are likely to play game-changing roles in fuelling these trends. The three main trends and their associated technological advancements are discussed as follows:

Mobility-on-demand is a change from an ownership model to a mobility-on-demand model that experts have unanimously agreed will play a key role in the future. Some mentioned that driving license applications are decreasing among the younger generation in some countries. This means that people are more willing to accept mobility-as-a-service instead of investing in cars as assets. The two technologies that are likely to accelerate this trend are shared mobility systems, especially car-sharing and ride-sharing apps, as well as autonomous vehicles.

Multi-modal transport, which is using different modes of transport to commute from one point to another, is the second most important trend. According to one expert from one of the leading automotive companies in the world, many global automotive companies are beginning to provide end-to-end mobility by partnering with different organisations. The key technology for this system to work seamlessly is **real-time information** availability of all transport modes, and the ability of commuters to interact with these modes on demand.

E-commerce has been around for some time now, but experts say that it is growing at a significant rate in Singapore. This trend is linked to the increased use of smartphones, which has allowed people to go online to shop more frequently. E-commerce will not only disrupt how people shop, but also how people will work and socialise.

4 Final Remarks

In this paper, we introduced foresight studies and the importance of adopting such tools in planning for the future of urban mobility in Singapore. In this ongoing study, we have discovered that trends in mobility-on-demand, multi-modal transport, and e-commerce will influence the way that transportation will evolve in the future. These trends, when twinned with challenges in demographics, culture, regulation, and city logistics, will help to develop a framework to visualise different scenarios that Singapore might face in the future. Our next tasks involve conducting more expert interviews, preparing a technology roadmap to identify possible game-changing technologies in urban transport, and preparing and analysing scenarios to discuss and evaluate them at a scenario planning workshop.

Acknowledgments This material is based on research/work supported by the Singapore Ministry of National Development and National Research Foundation under L2 NIC Award No. L2 NIC TDF1-2014-1. In addition, we would like to thank the Land Transport Authority for their inputs for this study, as well as the numerous experts from academia and industry for sharing their insights with us. The authors alone are responsible for the content and views in this publication.

References

1. Auvinen, H., Tuominen, A., Ahlqvis, T.: Towards long-term foresight for transport: envisioning the Finnish transport system in 2100. Foresight **14**, 191–206 (2012)
2. DHL: Delivering Tomorrow: Logistics 2050, A Scenario Study. Deutsche Post AG, Headquarters, Bonn (2012)
3. Ecola, L., Zmud, J., Gu, K., Phleps, P., Feige, I.: The Future of Mobility: Scenarios for China in 2030. RAND Corporation, Santa Monica (2015)
4. Edgar, B., Alänge, S.: Scenario planning—the future now. In: Alänge, S., Lundqvist, M. (eds.) Sustainable Business Development: Frameworks for Idea Evaluation and Cases of Realized Ideas, pp. 70–86. Chalmers University Press, Gothenburg (2014)
5. Fishman, T.: Digital-Age Transportation: The Future of Urban Mobility. Deloitte University Press, Westlake (2013)
6. Gazibara, I., Goodman, J., Madden, P.: Megacities on the Move. Forum for the Future, London (2010)
7. Hejazi, A.: Answering 18 Hot Questions on Forecast & Foresight (2011). http://www.academia.edu/1841949/Answering_18_Hot_Questions_on_Forecast_and_Foresight. Accessed 30 Aug 2015
8. IPS: IPS-Nathan Lectures (2014). http://lkyspp.nus.edu.sg/ips/events/ips-nathan-lectures. Accessed 30 Aug 2015
9. LTA: A World Class Land Transport System. Land Transport Authority, Singapore (1996)
10. LTA: Land Transport Masterplan. Land Transport Authority, Singapore (2008)
11. LTA: Land Transport Master Plan (2013). https://www.lta.gov.sg/content/dam/ltaweb/corp/PublicationsResearch/files/ReportNewsletter/LTMP2013Report.pdf. Accessed 30 Aug 2015
12. Lyons, G., Davidson, C., Forster, T., Sage, I., McSaveney, J., MacDonald, E., et al.: Future Demand. Ministry of Transport, New Zealand, Wellington (2014)
13. Martins, P., Boaventura, J., Fischmann, A., Costa, B., Spers, R.: Scenarios for the Brazilian Road Freight Transport Industry. Foresight **14**, 207–224 (2012)

14. MND: Land Use Plan to Support Singapore's Future Population (2013). http://www.mnd.gov.
    sg/landuseplan/e-book/index.html. Accessed 30 Aug 2015
15. MTI: New Challenges, Fresh Goals—Towards a Dynamic Global City (2003). https://www.
    mti.gov.sg/ResearchRoom/Documents/app.mti.gov.sg/data/pages/507/doc/1%20ERC_Main_
    Committee.pdf. Accessed 30 Aug 2015
16. NPTD: A Sustainable Population for a Dynamic Singapore (2013). http://population.sg/
    whitepaper/#.VeZ5J_aqpBc. Accessed 30 Aug 2015
17. Office of Science and Technology: Intelligent Infrastructure Futures. Department of Trade and
    Industry, UK, London (2006)
18. Popper, R.: Mapping Foresight: Revealing How Europe and Other World Regions Navigate
    into the Future. European Commission, EFMN, Luxembourg (2009)
19. Sim, J.: Beyond 50: Re-Imagining Singapore. Really Good Books Publishing House Pte Ltd,
    Singapore (2015)
20. Townsend, A.: RE-PROGRAMMING MOBILITY: The Digital Transformation of
    Transportation in the United States. New York: Rudin Center for Transportation Policy &
    Management Robert F. Wagner Graduate School of Public Service, New York University
    (2014)
21. URA: Concept Plan 1991 (1991). https://www.ura.gov.sg/uol/~/media/User%20Defined/
    URA%20Online/publications/research-resources/plans-reports/ltnl.ashx. Accessed 30 Aug
    2015
22. URA: Concept Plan 2001 (2001). https://www.ura.gov.sg/uol/~/media/User%20Defined/
    URA%20Online/publications/research-resources/plans-reports/concept_plan_2001.ashx.
    Accessed 30 Aug 2015
23. URA: Concept Plan 2011 and MND's Land Use Plan (2011). https://www.ura.gov.sg/uol/
    concept-plan.aspx?p1=View-Concept-Plan&p2=Land-Use-Plan-2013. Accessed 30 Aug 2015
24. van Voorst tot Voorst, M.-P., Hoogerwerf, R.: Tomorrow's Transport Starts Today. The
    Hague: STT Netherlands Study Centre for Technology Trends (2014)
25. Zhao, J., Liu, J., Hickman, R., Banister, D.: Visioning and Backcasting for Transport in Jinan.
    Transport Studies Unit, University of Oxford, Oxford (2012)
26. Zmud, J., Ecola, L., Phleps, P., Feige, I.: The Future of Mobility: Scenarios for the US in 2030.
    RAND Corporation, Santa Monica (2013)

Managing the Embedded Systems Development Process with Product LifeCycle Management

Eliane Fourgeau, Emilio Gomez and Michel Hagege

Abstract In industries like Transportation, Aerospace and Defense or High-Tech, the effective development of complex systems implies mastering multiple disciplincs and processes: project, people and data governance processes, requirements based and model based system engineering approaches, System Architecture Design with trade-off analysis, cross-disciplinary detailed design, integration, test, validation and verification methods. These processes are interrelated; managing complexity calls for tight collaboration between development stakeholders and demand very high data consistency between system compositional elements, at all levels. Organizations face significant challenges to maintaining coherency, consistency and relevance of this information across domains as well as capitalizing on it from one complex project to another; Failure to do so though is often resulting in costly reworks, product recalls, worse: attractiveness and competitiveness drops and jeopardized growth. In today's highly competitive, fast changing and demanding markets, saving investment time, resources and money for the creation of higher market appeal products is vital. This presentation will outline a vision for an open and integrated System Engineering platform, that leverages the richness and efficiency of online communities, the effective composition and reuse of multi-disciplinarily assets and the value of virtual user experiences, to facilitate new smart E/E systems development. Architecture of Systems is therefore just a unified and homogenous framework that allows to see all these different architectural traditions as instances of a same and unique "abstract" discipline. This system-oriented discipline clearly emerges nowadays from the convergence that one can presently observe at the level of the scientific and methodological foundations of many architectural practices such as: Enterprise Architecture that allows to design integrated enterprise information systems, Software Architecture which is at the basis

E. Fourgeau · E. Gomez (✉) · M. Hagege
Dassault Systèmes, 10 Rue Marcel Dassault, 78946 Vélizy-Villacoublay, France
e-mail: Emilio.Gomez@3ds.com

E. Fourgeau
e-mail: Eliane.Fourgeau@3ds.com

M. Hagege
e-mail: Michel.Hagege@3DS.com

© Springer International Publishing Switzerland 2016
M.-A. Cardin et al. (eds.), *Complex Systems Design & Management Asia*,
Advances in Intelligent Systems and Computing 426,
DOI 10.1007/978-3-319-29643-2_11

of all software design processes, Systems Architecture that is key for designing efficiently complex industrial systems, resulting from software-hardware integration, Organization and Project Architecture which underlies agile development methods. Architecture of Systems is therefore just a unified and homogenous framework that allows seeing all these different architectural traditions as instances of a same and unique "abstract" discipline, which underlies agile development methods.

1 Introduction

For more than 20 years, manufacturers and suppliers of automobiles, buses, trains, planes, satellites, defense systems, industrial equipment, complex communication systems and more have relied on the power of 3D CAD solutions to help achieve concurrent 3D development, seamless integration and effective digital mock-up creation, at the same time as they implemented Product Lifecycle Management (PLM), establishing a single source of product data and product record and a collaborative repository providing CAD, manufacturing, and service departments a centralized and controlled environment to house product data (Fig. 1). In this context, product data generally refers to all the information associated to product specification, CAD models, drawings, test procedures, documentation, operating procedures, quality documents, compliance documents, assembly instructions, and service manuals.

Fig. 1 OEM's landscape

Simultaneously, in most industries, an explosive growth in electronics and networked computerized systems with more and more software has been developing and experts predict this growth in embedded systems complexity will continue unabated. This mounting sophistication is accompanied with increasingly focused customer demands for regional, customized products based on market segmentation worldwide, including new emerging markets in India, China, and South America. This is no easy task to deliver, given the ongoing trend to execute the design and production of products, manage complex product lines and product variants with reduced resources and numerous specialists, often located around the globe.

As a result, both OEMs and suppliers are under intense pressure to improve the efficiency and effectiveness of how they design and build products. OEMs must deliver what the consumer wants, suppliers what OEM requests, when and where they want it, at an acceptable price; to achieve this, they need to collaborate across all departments—and across continents—and create innovative products that meet their customer needs as well as comply with industry regulations.

Only a profound evolution of the industry's infrastructure, processes and core technologies can meet this demand for accelerated innovation and help manage multiple complex product lifecycles. Reduced time to delivery and reduced budgets permeate the competitive landscape calling for new solutions that support distributed product development with global collaboration between stakeholders, not only for 3D disciplines but also for the complete System Engineering including embedded systems development processes.

PLM must expand beyond physical data and manufacturing, and include everyone with a role in a product's life—from end consumers, to internal functions developers, functional architects, embedded systems developers, tests engineers, integration teams, manufacturing and post sales staffs.

Accelerating innovation with PLM 2.0.

Despite the hesitation and general desire of Embedded System engineers to be left alone by those telling them, "we're from corporate IT, and we're here to help," they usually hate wasting time and effort. After all, time is money, and due to the pervasive aspects of Embedded Systems, embedded system engineers are under tighter deadlines and more stringent constraints for higher quality, faster pace development cycles, and cost effectiveness:

- Accelerate innovation introduction and improve quality, in line with end users product experience and use cases expectations.
- Deliver environmentally compliant products that meet global market demands.
- Ensure delivery of projects within performance and timing guidelines.
- Enhance global collaboration at multiple sites and throughout supply chain.
- Master the entire Embedded System development workflow, including requirement gathering and authoring, functional architecture optimization, logical decomposition, trade-off implementation for Software and Hardware, unit tests and integration tests, documentation,... and more.

- Manage end-to-end software architecture and systems engineering processes, including functional analysis, logical systems design, multi-physics behavior modeling and simulation, configuration and change management.
- Integrate multiple domains and bring together contributors from diverse technical disciplines to provide a high-level, comprehensive definition of a product.
- Validate the designed Systems as early as possible in the development process, when changes are least expensive to make.
- Produce the safest, high-performance systems with comprehensive requirements traceability.
- Ensure systems optimization including Software architecture, Hardware and Communications optimization.
- Manage engineering change order effectively.
- Reuse design/production data and preserve methods and processes across projects/programs.
- Provide accurate, real-time reporting of program status.

Why then should they care about PLM?

Of what value is a PLM based development paradigm?

How should this PLM based approach look like to contribute a great impact on embedded system engineering productivity, quality of designs, and time to market?

To answer the above questions, one should look at some aspects of today's reality for engineering communities:

- What about spending days trying to find development artifacts (requirements, functions, models, code, tests, documents, etc....) previously used in a project and finally giving up and redesigning new ones?
- What about trying to push for module based developments to leverage reuse while consistent data consolidation at module level is almost impossible?
- What about searching for the latest test results or the contracted product requirements from the customer, or trying to consolidate all tests run to validate the given function on all HW and SW product parts involved to fulfill the function?
- What about tracing requirements throughout all development artifacts data, tracking engineering change orders across the development cycle and analyzing change impacts on the different development teams' activities? Even though many companies use a formal change process to control changes to product data, as many engineers know, change processes aren't always as optimized or efficient as they could or should be, especially when they have to manage data changes across various heterogeneous files authored by as many independent, non-interoperable authoring tools. Some engineering change processes have become so bloated and time consuming that the change process itself becomes more arduous than the design process. How many engineers are still walking the engineering change through to each approver to ensure it gets priority attention? How many times are embedded systems engineers and others downstream, such as test engineers, waiting for the change before they can complete their tasks, address a quality issue, or respond to a new customer request?

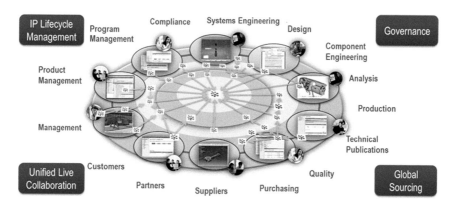

Fig. 2 PLM actors

Lack of real-time, comprehensive, efficient and accurate access to product data and customer/end user expectations has a great impact on engineering productivity, quality of designs, and time to market.

To achieve these goals, increase productivity, meet timelines and innovate efficiently, companies must master disparate sources of globally distributed intellectual assets and facilitate open, frequent, streamlined collaboration across the enterprise and extended enterprise and into the marketplace.

The goal of a PLM based system engineering development approach is to provide the development teams, including fast growing embedded system teams, unambiguous and easy access to all product and product variants data at its latest release level, to ensure on one side that everyone is sharing and is working with the most up-to-date information, on the other side that enterprise knowledge is capitalized across engineering departments, structured in generic modules for reuse in different projects as much as possible (Fig. 2).

Hence, there is a need for a single platform that consolidates product-related information, throughout the development process steps, making it visible and actionable by participants from all areas of the enterprise and when relevant of the extended enterprise. Powered by a PLM of second generation "PLM online for all", this platform must allow users everywhere to work concurrently in real-time on the same data, via a simple web connection to a single server.

Product data is managed through a series of defined lifecycle states, baselines, configuration and release levels. User access and IP are secured and controlled according to user, group, and role levels, from everywhere, at every time. Finally, the platform provides tracking and reporting tools to monitor the collaborative development effort progress. All these program management capabilities are woven into an easy to access environment of reduced risk and increased effectiveness.

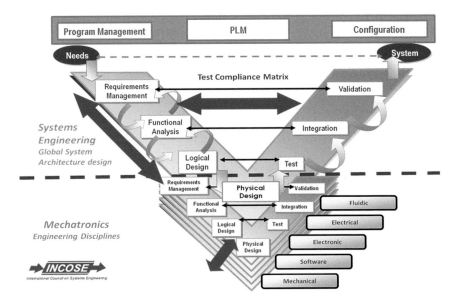

Fig. 3 PLM with MBSE scope

2 RFLP Data Model for Forward Looking Data Composition

The development of embedded systems involves multiple processes and disciplines (Fig. 3)—from project management and requirements engineering, through to configuration, mechatronics, HW and SW development, integration, test and verification management. The inter-disciplinary aspect of E/E and mechatronics development remains a challenge to all product developers, and especially for the sophisticated systems found in our today's life, making tight collaboration among the different design specialists mandatory.

These processes are interrelated and lead to a huge amount of systems data and models. Organizations face significant challenges and costs maintaining the accuracy and consistency of this information across all domains—with failure to do so often resulting in costly rework or even product recalls (Fig. 4).

To support effective embedded system development, the PLM based development approach calls for an innovative data model supporting seamless composition and easy reuse of complex product data, encompassing for a given product, system or sub-system, requirements and specifications documents, functions and functional interaction chains, logical compositions featuring HW and software trade-off architectures, eventually 3D Physical data and executable SW.

This is the so called RFLP (Requirement, Functional, Logical, and Physical) data model which in essence, helps guide the embedded system engineering teams chartered with the overall design (Fig. 5).

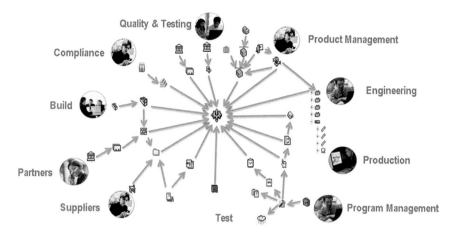

Fig. 4 Actors around PLM component

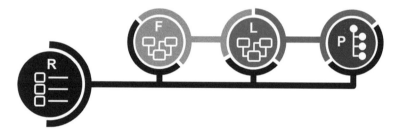

Fig. 5 RFLP framework

The Requirement and tests (R) defines customer and stakeholders needs and as the results of their analysis in technical terms: what characteristics, activities the system shall satisfy. The test defines the requirement verification protocol, the acceptance criteria, and the results.

The Functional (F) defines the mission that the system shall perform, what the system does, what actions or activities that must be accomplished to achieve a desired outcome, or to provide desired capabilities. The Functional formalizes the understanding of the intended use of the system (operational functions, use case scenarios, services) and its decomposition into elementary functions that can be allocated to technical solutions.

The Logical (L), or logical/organic defines how the system is, the technical solution based on subsystems, components and their interrelations according to stakeholders/customer needs and requirements, technical requirements, and expected functions.

The Physical View (P) defines a virtual definition of the real world product, including 3D representation to "visualize" the targeted system concept in the early phases.

The Functional and Logical include behavior models to define how functions or logical components transform inputs into outputs, react to events, or respond to excitations.

The Process and Methodology approach necessitates capturing the "voice of the customer" requirements and then applies them to make system level decisions downstream the development cycle, at the functional architecture level. These systems then get instantiated through subsystem-level functional structures and the logical systems that control them. Individual designers can then model and simulate these complex structures and the connections between them. Finally they are realized by actual physical components, whether mechanical, electrical, or software controls.

Functional and logical structures that have been defined between requirements and the physical components produced to satisfy those requirements are thus captured and aggregated in the same RFLP data structure.

These intermediate structures allow designers to analyze, model, simulate and virtually "Experience" the system and validate that the requirements are met before significant resources are engaged to its detailed design, development, validation and operation.

The structures produced by this RFLP based process allow them to understand, improve and capitalize the design for their reuse in new products developments.

3 Requirements and Model Based Engineering for Robust Process

3.1 Requirement Based Engineering

Taking into account the current and well established practices of various OEMs and Suppliers, effective support of Requirement based engineering processes means capturing requirements from many files and database sources (Windows/Office, Doors, ALMs, ReqIF,...) in a wide variety of data and file formats and link them (with easy to implement traceability links) to development and verification artifacts, through the provision of a large number of interfaces to common systems engineering tools. These traceability links should not only operate at files level, but truly at fine-grained engineering artifacts levels, enabling effective requirement coverage and change impact analysis.

Key benefits are:

- Capture of requirements at any level
- Intuitive, quick and easy requirement capture from various existing sources (Doors, Word or PDF documents,...)
- Coverage analysis and traceability
- Management of requirement changes and lifecycle information
- Upstream and downstream impact analysis for regression risk management
- Automated report and documentation generation.

3.2 Model Based Design

The typical E/E design process usually begins by generating specifications to describe the system, subsystem, component, and interface requirements. The system design is partitioned into functions and the functions partitioned into hardware and software. During development, the hardware and software components are usually developed individually, and then brought together late in the development cycle, when the system was integrated. In most cases, the final integration step, when discrepancies between concept and implementation are discovered, turns out to be the schedule driver: issues have to be resolved that were unforeseen, and thus not planned for!

Since recently, a novel methodology is emerging that incorporates the best features of the typical development cycle and adds additional visibility and risk reduction to the schedule-driving integration phase. This is accomplished by using, early in the design cycle, advanced hybrid simulation techniques that allow a lifelike, "virtual" integration of multi-faceted component models that can be launched while the design is still in its early phase and easy to change. This virtual integration eliminates the costly reworks that are frequently required when problems are not discovered until late in the product integration and validation phase. It also drastically reduces the time spent integrating hardware and software, sensors, actuators, processing units and communication networks, because many of the problems have already been discovered and corrected during development.

Each element at each level of hierarchy will have its own development flow, which will resemble the development flow of the overall system. Eventually many interoperable models will have to be developed to support the various operational behaviors that need to be simulated and validated (finite element model, multi-physics/mechatronics dynamic model, dysfunctional model for a safety behavior assessment, thermal model, control model, etc....) (Fig. 6), ultimately with cross-correlation parameters reflecting impact of 3D metrics on behavior dynamics for instance, or of logic trade space effects on physical dimensioning and vice et versa, or of non-functional constraints on COTS selection, etc....

It is useless to stress that complete systems depend on software (as well as hardware), and that system software design can make or break any system in performance, schedule, and cost. Software architecture and performance can be modeled just as hardware can. Real time simulations, hybrid simulation techniques such as MIL, HIL and SIL are to be used, as early as possible in the development cycle, with the capability to interoperate with other simulation techniques, which are necessary to virtually assess the complete system operational behavior.

Key aspects of a real system can be modeled and simulated completely in a lifelike simulation of complex multi-disciplinary systems, addressing the known characteristics of the system that relate to its multi-dimensional behavior and performance with regards to specific real life use cases and environmental conditions. This is known as the lifelike digital experience paradigm, where functional

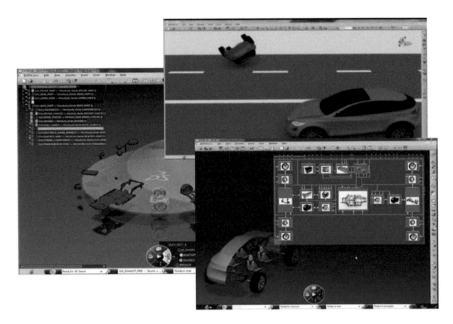

Fig. 6 3DEXPERIENCE platform

mock-up meets digital mock-up, both in one, to virtually express a user experience and reason on it from different Engineering perspectives.

This trend is nurtured by novel modeling and co-simulation technologies.

The industry has been striving for years to provide a unified open standard language (Fig. 7) to describe multi-physical systems behaviors. Eventually such a language, the Modelica language, has been developed and is starting to enjoy a successful adoption by the E/E mechatronics community worldwide.

Furthermore, to support Model based Design, the emerging FMI standard brings multiple interoperable simulation capabilities: the Functional Mock-up Interface (or FMI) defines a standardized interface to be used in computer simulations. This interface is implemented by simulation models or software libraries called FMU

Fig. 7 Openness axes

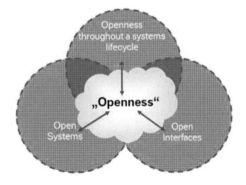

(Functional Mock-up Unit). FMI development was initiated by Daimler AG and is developed by the MODELISAR consortium. FMI is the runtime environment that was created to allow simulation data exchange (see also http://functional-mockup-interface.org/). By supporting the open source Modelica language, and leveraging the FMI interoperability of models and simulation kernels (co-simulation), The PLM based platform will also help Systems Engineers execute and analyze system and sub-systems models, while mixing dynamic and state logic behaviors, adding control by linking to Matlab Simulink or other control models. Modelica makes it possible for users to generate deterministic behavior and reuse modular mechatronics components. The open Modelica Standard Library, which contains mechanics, thermal, electricity including electronics, fluidics and control components, greatly improves collaboration across engineers of various disciplines, as well as enhances modeling effectiveness and throughput.

Built-in openness for scalable adoption.

Requirement based Engineering and Model based Design are two main pillars of effective system engineering processes. To support them, built-in openness mechanisms have to be implemented in the platform, with manifold objectives:

- Scale requirement traceability and change impact analysis throughout most popular systems engineering tools
- Support E/E and mechatronics industry standards.
- Integrate or interoperate with complex and heterogeneous System Engineering and Simulation environments by using standards as FMI.
- Be open for partners' solutions including customer preferred partners.
  Enable controlled and stable integration to other systems such as PDM, ERP, xCAD using the xPDM and xCAD platforms.

4 Conclusion

Market forces are driving the quantity, quality, completeness, and rate of change of Embedded System Engineering environments. Primarily, better integration and traceability between product requirements and lower level technical requirements down to system/software engineering environments are called for. An immediate powerful result of this integration will be the comprehensive traceability between high level driving product/modules requirements, the resultant technical configurations, and their cost and schedule impacts (design-to-cost, just-in-time ...).

The requirement to integrate the trade-off space across technical, program and financial boundaries is likely to grow up. Models authoring will populate rich reusable model libraries if these engineering environments support open, standard languages and description formats, fostering their adoption and ease of inter-operation by a large community of domain skilled engineers worldwide.

These collective, multi-disciplinary efforts would never be harnesses nor capitalized upon without a true open, multidisciplinary, collaborative system engineering management platform that facilitates comprehensive models elaboration, interoperation, cross correlations, global parameterization, configuration and reuse.

To accelerate the development of innovative products with higher rates of market success, online, collaborative PLM based solutions are mandated for both OEMs and Suppliers.

Such an open and integrated E/E PLM portfolio will enhance business processes and facilitate global collaboration, transforming information into actionable and reusable knowledge in a lifelike digital experience and effectively manage strategic activities, all while collaborating with global teams and key decision makers.

References

1. Dr. Morkevičius, A.: Integrated Modeling
2. Henneau, P.: A pragmatic approach to SE practices at Schindler elevators
3. Krob, D.: Architecture des systèmes complexes
4. Krob, D.: Eléments d'architecture des systèmes complexes
5. Krob, D.: Eléments de systémique—Architecture des systèmes 1
6. Resser, A.: Model based systems engineering management plan
7. V-Model Views, INCOSE

Establishment of Model-Based Systems Engineering Approach for Rapid Transit System Operation Analysis

Mohamad Azman Bin Othman, Tan Chuan Heng and Oh Sin Hin

Abstract Model-Based Systems Engineering (MBSE) for Rapid Transit System (RTS) applications has recently been adopted by several established railway system suppliers. However, the application of MBSE at the System-of-Systems (SoS) level remains less prevalent in the transportation industry. The adoption of MBSE using Systems Modelling Language (SysML) in this area would offers better management of SoS complexity, enhanced communications among stakeholders and greater support for RTS operators leveraging on models across the systems' lifecycle.

1 Introduction

Designing and building a modern and driverless Mass Rapid Transit (MRT) system has become increasingly challenging today as it involves a huge number of supporting systems across multiple engineering domains (mechanical, electrical, civil and environmental), tightly coupled to one another. Figure 1 shows a typical MRT system, expressed in a top level view together with its main supporting systems.

The successful operation of the MRT system depends heavily on the proper execution of these systems' functionalities at an individual level, as well as at an integrated level. These systems are intricately intertwined so as to achieve the desired emergent properties and performance that consistently meet high standards of reliability, availability and safety.

M.A.B. Othman (✉) · T.C. Heng · O.S. Hin
Land Transport Authority, 1 Hampshire Road, Singapore 219438, Singapore
e-mail: Mohamad_Azman_Othman@lta.gov.sg

T.C. Heng
e-mail: Tan_Chuan_Heng@lta.gov.sg

O.S. Hin
e-mail: Oh_Sin_Hin@lta.gov.sg

© Springer International Publishing Switzerland 2016
M.-A. Cardin et al. (eds.), *Complex Systems Design & Management Asia*,
Advances in Intelligent Systems and Computing 426,
DOI 10.1007/978-3-319-29643-2_12

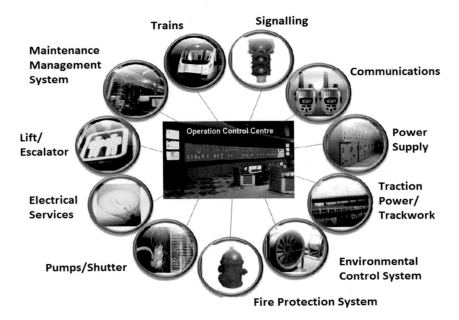

Fig. 1 The composition of a modern MRT

However, the involvement and integration of such multi-disciplinary systems increase the design complexity of a MRT system. Project teams managing the construction of a MRT system not only have to closely monitor the development and testing of individual supporting system, they will also need to understand the design and behaviour of the MRT system as a whole, when these supporting systems are integrated to achieve its desired operation needs.

In addition, due to increasing demand for the expansion of transportation infrastructure, project teams face many challenges to explore new methodology in order to optimise the time required for the design and delivery of a new MRT system. Any ambiguity or uncertainty during the design phase may result in increased costs and prolonged project delivery.

In the past, document-based approach was employed for previous RTS projects. This approach involved manual requirements elicitation, creating drawings manually and independently, producing static diagrams with stored views and processes that depends on manual checking and updating to ensure consistency. This resulted in much difficulty when managing a complex system such as a MRT system.

To overcome such uncertainties, MBSE has been identified as a suitable approach that the project teams can adopt to manage the enormous complexities in their work, especially in the Systems Integration (SI) domain. In addition, with the utilisation of MBSE, the following objectives could be realised:

- To enhance communications among stakeholders (operators, contractors, regulators, asset owner, external agencies and passengers)

- To improve quality of work
- To increase work productivity
- To reduce development risks

In this paper, formulating an approach towards implementing MBSE for automated driverless MRT projects is shared.

1.1 Implementation Consideration

Through the utilisation of SysML, models and diagrams are constructed to provide an accurate and coherent view on the system design and operation of a RTS. System requirements could be elicited, analysed and decomposed to lower level details, facilitating the partitioning and allocation of functions to the respective stakeholders. Also, models that are developed can be linked and traced back to their requirements, ensuring consistency, completeness and correctness of the system design.

Using a RTS project as a reference, a tool based on SysML (IBM Rational Rhapsody®) has been explored to implement models representing various aspects of the RTS, such as operations concept, SoS architecture and operational scenarios depicting various operating conditions.

However, recognising that individual system modeller has different techniques and preferences, there is a possibility that these diversities in design styles adopted by the system modellers might result in the inconsistency of the system models at later stages. The immediate challenge is to align the design in a standard development approach, thus ensuring the system models (and diagrams) remain clear and consistent to other stakeholders. To address these concerns, the issue was looked into and a proposal to adopt a common MBSE framework was established.

The common MBSE framework governs the approach and guides the system modellers on the techniques to model a RTS project. In addition, a supporting document will be compiled so as to dictate the general guidelines and best practices in system modelling, and will serve as a reference for other project teams.

2 Approach in Implementing MBSE

The main objective for establishing a common MBSE framework is to develop a common baseline profile based on standard guidelines and best practices in order to ensure clarity and consistency of models and diagrams, enhancing reusability and concurrent models development. System resiliency checks and analytical studies can be easily defined within the context of the common framework, without any modification to the system models.

With the MBSE framework, two outcomes were sought after. First, the framework layout (embedded with the baseline profile) should be simple and concise, so that any third party would be able to easily access the models through the package descriptions and understand the whole or part of the RTS model based on their interest. Second, a baseline profile will be established in Rational Rhapsody® environment. This profile could then be loaded into individual system modeller's workstation before the start of the project. This will ensure that the start off is from a common baseline and there would be no deviations in the model tool settings.

2.1   MBSE Framework

With the above considerations in mind, the proposed MBSE framework is shown below in Fig. 2.

The highest level consists of Project_TEL_Model (thereafter defined as the SoS) to depict the modelling of SI works on Thomson-East Coast Line (TEL). The system model will be broken down further into eight sub-packages namely: *Structure, Behaviour, Use Cases, Requirements, Parametrics, Viewpoints, IO Definitions* and *Value Types*.

2.1.1   Operation Concept and System Requirements

Firstly, use cases were used to define the SoS and its boundary. Use case diagrams were used to develop models that illustrate the possible operational scenarios that the SoS can be used. These use case diagrams are consolidated under the *Use Cases*

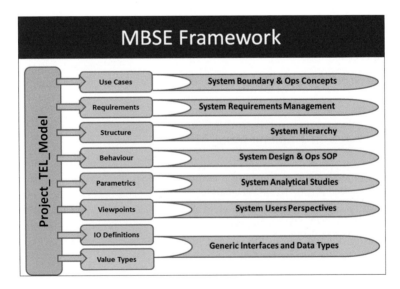

Fig. 2 MBSE framework

package to facilitate better readability from stakeholders who are keen to understand the operational concepts only, without going through the technical details. This will be further elaborated in Sect. 3.

Next, the source requirements associated with the SoS were established by using another tool (IBM Rational DOORS®) for the management of requirements. These source requirements are loaded into Rational Rhapsody®, under the *Requirements* package to fulfill traceability with use cases. From then on, various aspects of the system design were worked concurrently; such as identifying the relevant system functions to satisfy the requirements, formulating the SoS architecture, establishing the functional components and their interfaces.

2.1.2 System Design and Analysis

Block definition diagrams are utilised to illustrate the overview of the SoS, in which the system components are represented as blocks in these diagrams. Detailed design of the system components are illustrated using internal block diagrams, showing the ports and inter-relationships between the components. These diagrams will be grouped under the *Structure* package, which stakeholders can access the contents easily and understand the architecture, hierarchies and their interconnections without much navigation within the SoS.

A series of diagrams (activity, sequence, state machine) was used to illustrate actions, message exchanges, state transitions and events by the stakeholders and/or systems. As these diagrams represent the behaviours of the SoS, they are jointly consolidated under the *Behaviours* package. This package was produced to provide the functional breakdown of the SoS and their detailed message interactions and timings between the system components.

SoS parameters and constraints are illustrated by parametric diagrams in SysML and they are grouped under the *Parametrics* package. The package also consolidates various analytical initiatives, such as performance studies, impact analysis, and design trade-off studies. By accessing the *Parametrics* package, the ability to assess the performance aspects, regulate and monitor system variables without affecting the core design of the SoS is realised.

2.1.3 Stakeholders' View on the System

Individual stakeholder has a different perspective view of the SoS, for example the Station Operator will be more focused on the station aspect as compared to the Traffic Control Operator who will be concerned on train regulation at the Operations Control Centre (OCC). The framework consists of a *Viewpoints* package, with the purpose of framing these perspective views into a common location. This package will contain all views from stakeholders in their respective area of responsibilities (OCC/Train/Station/Depot) and their "linked" activities.

Fig. 3 Baseline profile in rational rhapsody®

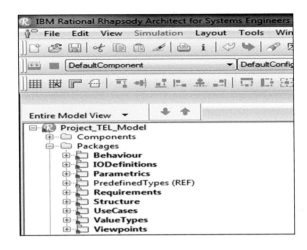

2.1.4 Generic System Interfaces

The *IO Definitions* and the *Value Types* packages contain generic system interfaces and data types that are typically used in RTS projects. These are defined based on international standards, in-house engineering standards, best practices and past projects' references. Through these packages, generic models packages are built and these models can be re-used in future RTS projects.

System models developed will be categorised under these sub-packages respectively. In this manner, the simplified layout allows the re-usability, readability and concurrent models development to take place. The baseline profile is developed using the Rational Rhapsody® tool, as shown in Fig. 3.

In parallel to the MBSE Framework, a list of guidelines will be defined to assist the system modellers in their design work. This is to ensure that their works are aligned and maintained at a high level of consistency, thus improving the readability of the system models. As such, a document, *Best Practices of SysML Modelling* was created with the following objectives:

- To provide general system modelling design principles
- To adopt common/well-known terminology in RTS domain
- To adopt a standardised layout convention for various models and diagrams
- To adopt a standardised colour convention for various model elements.

3 Operational Analysis Using MBSE Approach

Adopting MBSE in analysing the rail operation requirement provides benefits that cannot be achieved through conventional document-based approach. The MBSE approach requires the stakeholders involved to be identified and their relationships to be defined in the early stage of development. This results in a more specific and accurate outcome of the study performed, which then translates to a RTS operation model that truly reflects its intended purpose.

Using various SysML diagrams, a clear representation of the relationships and behaviours of the SoS can be portrayed effectively. Within the same project, a single system component declared and used in different diagrams will ensure that traceability can be viewed in the model throughout the system life cycle. Also, this presents an opportunity for impact and trade-off analysis to be performed as any modification to that component will generate a corresponding record of repercussions in the model.

The ability to simulate the operation models offers a glimpse of how the actual system will behave under stressed (e.g. maximum capacity) or degraded conditions (e.g. component failure) and thus exposing inherent system vulnerabilities. In this way, it checks the resiliency of the SoS and proposes improvements such as preventive or corrective mitigations to improve the robustness of the RTS.

During the concept stage, identifying the stakeholders and their needs is the priority in taking a step in the right direction. The high level stakeholders' needs and relationships can be represented using a use case diagram as shown in Fig. 4. The stakeholders are identified as "Actors" while the packages representing the operation regions are placed inside the RTS boundary box. Use cases are placed at appropriate regions to represent stakeholder's needs and association.

Proceeding to design stage, the stakeholders' needs can be further classified into second level behaviour or activities required by the SoS during normal, degraded and emergency mode of operation as shown in Fig. 5. The stakeholders, in particular the operator, are also expanded further to elaborate the actual positions or designations in the organisation that have direct interactions with the SoS.

By analysing a specific scenario for normal, degraded or emergency mode of operation, the actions required by different "Actors" can be described for the normal day-to-day activity or to return the SoS to normal state when faced with an unexpected scenario. Using activity diagrams (Fig. 6), all involved "Actors" are arranged horizontally in a swim lane diagram. The activity box in that lane indicates the action required by the particular "Actor" to perform or to decide at a certain point of the operation scenario. This will provide clear responsibilities on who should do what, when the need arises or how the SoS should behave in different scenarios.

The activity diagram has provision "call behaviour" for a nested scenario to link up related scenarios when certain conditions warrant for it. Clicking on this box will invoke the function to call up the appropriate scenario. This avoids repetitive creation of the same set of activities from appearing in different diagrams.

Notice that in Fig. 7, the activity boxes are "linked" to various functions of the SoS. These functions are identified by their unique function numbers (e.g. FO2.2.1). These linkages from the activity boxes are required for the down tracing of required functions that are supporting the activity. These activity boxes can also be hyperlinked to the related function for linking purposes and ease of viewing.

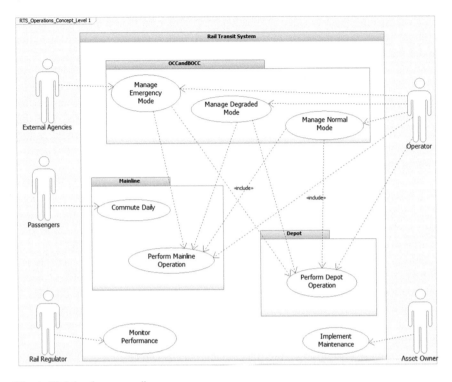

Fig. 4 High level use case diagram

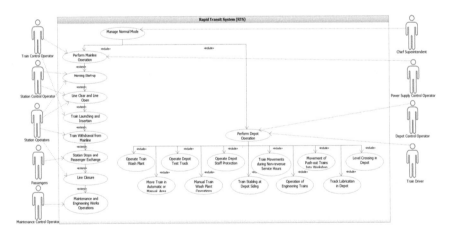

Fig. 5 Second level behaviour use case diagram

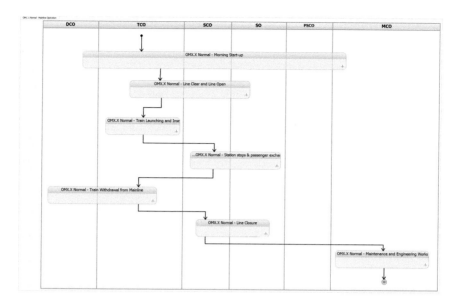

Fig. 6 Representing operational scenario using activity diagram

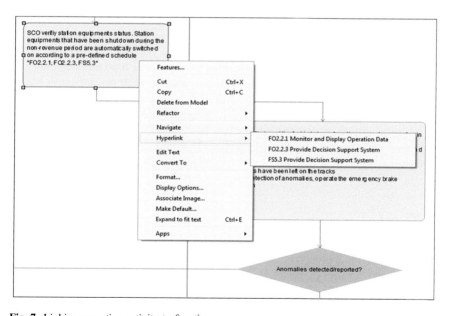

Fig. 7 Linking operation activity to functions

4 Model-Based Approach Versus Document-Based Approach

The model-based approach was adopted for a new RTS project, instead of the previous document-based approach. At this juncture, comparison between the two approaches is made to reflect on the improvements achieved with the new approach.

Firstly, using the model-based approach, missing links were able to be identified in design documents submitted by different contractors. Utilising the traceability features provided by the modelling tool, these missing links were uncovered. The cause is due to the usage of multiple link references in various documents and the eventual removal of a base referenced diagram. This problem is prevalent in document-based approach, as reviews of each document could have been done at a different time and there was unawareness on the changes made to the source reference. Maintaining a proper record on the links could be a solution to this but much time would be incurred on updating this record manually.

Secondly, system components that were critical to the SoS were easily identified. Through the system models, information such as systems dependency, quantity and safety-critical connections can be easily derived and represented in a single diagram or chart. Also, there is an additional advantage to develop impact assessment and performance analysis, simply by defining some simulated scenarios based on these system models. This would have taken substantial efforts in the document-based approach, especially when this information exists across several design documents.

Lastly, with the system models created, there is a possibility of reusing these models for future RTS projects. By extracting a base model from existing projects, this can be utilised as a starting point for a new project. With the document-based approach, significant efforts are often incurred in changing the state chart and architecture diagrams manually. This is unproductive and inefficient. Using model-based approach, system models can be reproduced and updated easily to create design documents for the new project. The reduction in time allows system engineers to focus their attention more on system design and analysis.

5 Conclusion

Clearly the MBSE approach offers clarity, traceability, re-usability and coherency of the SoS design and development. Recognising the potential challenges, an approach which consists of the MBSE framework and best practices to assist the system modellers in their area of work was formulated. Through this established approach, the aim is to align the development direction among the system modellers, thus moving towards the ultimate goal of managing system complexities at a SoS level.

Also, a simple comparison was performed between model-based approach and document-based approach. Even though significant efforts were spent on developing the system models, the potential benefits from the model-based approach far outweighs the costs. Exciting possibilities are achievable with the new approach and improvements to productivity level can be expected.

Moving forward, there will be continuous efforts to develop models representing system functions supporting the RTS. The plan is to provide an exhaustive functional analysis model depicting the required high level to lower level functions. Individual function will be allocated to each system for realisation. Subsequently, analysis on the interface or data exchanges between these systems can be carried out by producing a data flow model. Amongst other things, this model should indicate the type of connection (e.g. LAN or hardwired), the type of information (e.g. control or data) and the direction flow of data (i.e. origin to destination). Also, the performance of the interface can be optimised by analysing the information exchanges. This analysis will be carried out progressively in subsequent stages in the future.

References

1. Friedenthal, S., Moore, A., Steiner, R.: A Practical Guide to SysML: The Systems Modeling Language (2012)
2. Hin, O.S., Joyce, H.P.F., Chan, S.: A practical implementation of a generic approach to systems engineering. In: INCOSE Conference (2010)
3. Hin, O.S., Joyce, H., Othman, M.A.: A systems thinking approach to building large-scale land transport systems. Mag. IES Singapore (2015)

A Model-Based Testing Process for Enhancing Structural Coverage in Functional Testing

Yanjun Sun, Gérard Memmi and Sylvie Vignes

Abstract Developing complex safety-critical systems usually involves developing models as abstractions in the upstream phases of design. It is still today often challenging to convince the industry that performing functional testing on models of systems may help reducing the cost of system testing. This article presents a new model-based testing process. Part of the "CONNEXION" French I&C methodology project, it combines a vast number of verification tools. In this article, we focus on the integration of a specification-based test generation tool, a model-checker and an environment for model test execution to enhance structural coverage rate. To this end, we define a novel process describing how to extend the functional test bed to enhance structural coverage by generating new test cases reaching so far uncovered branches using model-checking.

1 Introduction

In safety-critical industries (avionic, nuclear, automotive, etc.), systems are required to be developed under standards and certifications [1]. Apart from building the specification, upstream phases of design usually involve developing formal models

This work is founded by "CONNEXION" cluster (a project of French Investment in the Future Program) https://www.cluster-connexion.fr/.

Y. Sun (✉) · G. Memmi · S. Vignes
CNRS LTCI-UMR 5141, Télécom ParisTech, 46 Rue Barrault, 75013 Paris, France
e-mail: yanjun.sun@telecom-paristech.fr

G. Memmi
e-mail: gerard.memmi@telecom-paristech.fr

S. Vignes
e-mail: sylvie.vignesg@telecom-paristech.fr

© Springer International Publishing Switzerland 2016
M.-A. Cardin et al. (eds.), *Complex Systems Design & Management Asia*,
Advances in Intelligent Systems and Computing 426,
DOI 10.1007/978-3-319-29643-2_13

171

as an abstraction of system. It is cost-effective to perform testing on these models since the cost of bugs found later in the actual system can be extremely high.

In 2012, the main industrial and academic partners of French nuclear industry initiated a large R&D project called "CONNEXION" [2, 3]. One among several objectives of this project is to perform tool-supported automatable verification activities on the formal models built in early stages of I&C system life cycle. This falls into the category of model-based testing.

Traditional model-based testing [4, 5] is a variant of testing that relies on abstract formal models which encode the intended behaviour of a system. Test cases can be automatically generated from models, usually they will be executed on the system under test. In our study, test cases will be directly executed on executable models.

In this article, we present a new general tool-supported model-based testing process to be applied in "CONNEXION" with its specific set of verification tools. More precisely, our process combines a specification-based test generation tool, a model-checker and an environment supporting execution of test cases on models. This process aims at enhancing the structural coverage rate. The first tool generates test cases according to the formal specification to test functional aspects of models. With the help of the model-checker, we extend the functional test bed to enhance the structural coverage of models. This process is designed to address verification of vast portions of an I&C system.

The paper is outlined as follows: Sect. 2 presents the system engineering process in "CONNEXION", Sect. 3 summarizes some basic aspects of testing methodology; Sect. 4 presents a new process for enhancing coverage using a model-checker; Sect. 5 describes the environment built in "CONNEXION" that enables our process; Sect. 6 concludes and discusses future work.

2   Triple-V Cycle in "CONNEXION"

The objectives of the "CONNEXION" cluster concern Control Systems and Operational Technologies to maintain a high level of safety and to offer new services improving the efficiency and the effectiveness of operational activities. In the context of Nuclear Power Plants, an Instrumentation & Control (I&C) system is composed of several hundreds of **Elementary Systems (ES)**, controlling with a very high safety level thousands of remote controlled actuators: about 8000 binary signals and 4000 analog signals sent to the control room concerning over 10 000 I&C sub-functions and over 300 I&C cabinets. An ES is a set of circuits and components which perform an essential function to the operation of the nuclear plant. Each ES is divided into two parts: the **Process**, representing the physical infrastructure (heat exchangers, sluice gates, pipes, etc.) and the **I&C system**, responsible for the protection, control, and supervision of functioning of the process. The functional aspect of an elementary system is described by **Functional Diagram (FD)**, built in a non-executable formal language which complies with IEC standards [6, 7].

The development process of I&C system is illustrated by Fig. 1. The current approach corresponds to a V cycle including phases 1, 2, 4, 7, 8, 9, 10. The process starts with building a **specification of functional requirements** for both process and I&C of one ES. A more detailed functional specification of the I&C system is then deduced in form of a Functional Diagram. This Functional Diagram is progressively elaborated into a **Refined Functional Diagram (RFD)**, ready to be transformed to programs implemented on automata. The Verification and Validation (V&V) of I&C system is performed at the level of automata: first verification of I&C system of one ES with respect to its RFD; then verification of this I&C system integrated with I&C systems of other ES that are already validated. All the validated Elementary Systems are finally integrated through a platform to further validate the proper functioning of these automata in a simulated environment [2].

It has been proposed in "CONNEXION" to introduce two additional verification processes in the upstream phases of design: verification of FD and RFD with respect to functional requirements. This is defined as functional validation in [8]. The original V cycle is therefore enriched by two sub-cycles: 1, 2, 3 and 1, 2, 4, 5, 6. Due to the fact that FD and RFD are formally verifiable but not executable, these verification activities are currently manual. To automate the functional testing, the FD and RFD need to be equivalently recreated in an executable formal language. The partners of "CONNEXION" provides a complete tool chain to automate functional validation as much as possible. The development process is aligned with IEC standard [9], encouraging test automation tools. Our process proposes to use formal tools as test generators not as provers. Our process proposes to assist the manual test production phase but not to substitute it. In such a way, the integration of these tools in our process does not imply a challenging qualification of the tools. The current life cycle relies on a document-centric system engineering approach; "CONNEXION" enables the transition to a model-centric practice as advocated by the INCOSE [10]. The introduction of the two V sub-cycles answers for a key

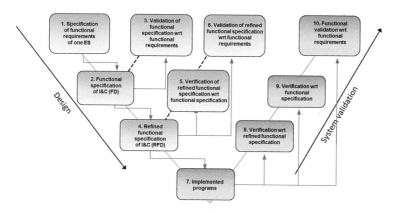

Fig. 1 I&C system life cycle in "CONNEXION"

challenge of Industrial Cyber-Physical Systems: early stage of Verification and Validation [11].

In this article we present a new testing process intended for the two sub-cycles of functional validation. It is designed to be applied first to a single ES and then to scale up to a cluster of ES.

3 Testing Methodology

3.1 Black-Box Testing and White-Box Testing

In general, there are two mainstream testing strategies: black-box testing (or functional testing) and white-box testing (or structural testing).

The goal of black-box testing is to find errors in the program by testing the functions listed in the specification. Designing test set does not require analyzing the details of code but using the specification. On the contrary, white-box testing focuses on the structure of the program. A test set in which every line of code is executed at least once is an example of white-box testing.

A mixture of different strategies can be used to improve the effectiveness of testing. For example, since black-box testing does not require knowledge of the structure of the program, there may be some parts of the program that are unreachable because they are defective or insensitive to certain inputs. These problems may not show up in functional testing.

The "CONNEXION" project looks for functional verification on models of system with respect to the specification of functional requirements. Meanwhile, covering as much of the model structure as possible (or at least some parts) is also an objective. This inspires us to design a testing process that combines black-box and white-box testing.

3.2 Coverage Criteria

Coverage criteria indicate how adequately the testing has been performed. According to the testing strategies presented just above, there are at least two categories of coverage criteria: requirement coverage and structural coverage. We hereby adopt the definitions of different coverage criteria given in [12].

A **requirement** is a testable statement of some functionality that the system must perform. **Requirement coverage** demands that all requirements are covered in the functional test set. In other words, a measurement of the requirements that are covered in the test set indicates how well the functional testing has been performed.

As for **structural coverage**, many criteria have been discussed in the literature. Statement coverage and branch coverage are two of the most widely used criteria in practice. A measurement of statements or branches covered in the test set indicates the test adequacy [13].

Definition 1 (*Statement coverage*) The test set must execute every reachable statement in the program.

Definition 2 (*Decision coverage* (*also called branch coverage*)) The test set must ensure that each reachable decision is made true and false at least once.

The testing of a decision depends on the structure of that decision in terms of conditions: a decision contains one or more conditions combined by logic operators (*and, or* and *not*). Several decision-oriented coverage criteria are hence derived:

Definition 3 (*Multiple condition coverage* (*MCC*)) A test set achieves MCC if it exercises all possible combinations of condition outcomes in each decision. This requires up to 2^N test cases for a decision with N conditions.

Definition 4 (*Modified condition/decision coverage* (*MC/DC*)) A test set achieves MC/DC when each condition in the program is forced to true and to false in a test case where that condition independently affects the outcome of the decision. A condition is shown to independently affect a decision's outcome by varying just that condition while holding fixed all other possible conditions. For a decision containing N conditions, a maximum of 2N test cases are required to reach MC/DC.

Indeed, these different structural coverage criteria are not equally strong. For example if we reach MCC then we reach MC/DC since MCC requires strictly more test cases than MC/DC. Similarly decision coverage is stronger than statement coverage. More detailed information about the hierarchy of coverage criteria can be found in [12].

Requirement coverage and structural coverage are somewhat independent in the sense that a 100 % requirement coverage does not guarantee a 100 % structural coverage and vice versa. In practice, the functional testing of large scale system is usually performed by executing a set of functional test cases in a harness. Often the statement coverage is considered as an indicator of testing effectiveness [14].

In this paper, we consider the branch coverage criterion, but the methodology can be applied to other structural coverage criteria with support of proper tools.

4 A Tool-Supported Model-Based Testing Process

The idea of coverage-based test generation using model-checking has already been studied by a few researchers. Geist et al. [15] proposes using a model-checker to generate a test case that covers certain areas of the program. Ratzaby et al. [16] uses model-checking to perform reachability analysis, i.e. whether certain areas can ever be covered by any test case. In [17], model-checking is used to derive test cases from all uncovered branches to enhance structural coverage. Geist et al. [15] and Ratzaby et al. [16] explains how one test case can be built while [17] presents a procedure for dealing with every uncovered branch. In this paper, we improve the procedure in [17] in particular by a refined termination test and by considering that a model-checker can have a "time-out" situation. In this last case, we propose an hybrid simulation using both a simulation and a model-checker as it is done for hardware design verification (see [18] for instance). Model-checker explores every possible execution whereas hybrid simulation explores only partially the set of all possible execution although driven by user inputs.

As to answer the question "how to design a test case using model-checking", [19] presents a framework where structural coverage criterion MC/DC is formalized as a temporal logic formula used to challenge a model-checker in order to find test cases. Also techniques for using model-checker to generate test cases from coverage criteria including branch coverage and MC/DC are described in [1].

Figure 2 illustrates our testing process and positions three tools that are requested to implement it. A model-based test generation tool is first used to derive a functional test set (*TS*) according to the specification (step 1). This test set is then executed in a proper environment on the model of system (step 2). We suppose that the requirement coverage (*RC*) and structural coverage (*SC*) of the test set are measured after the execution (step 4). Another assumption is that the information

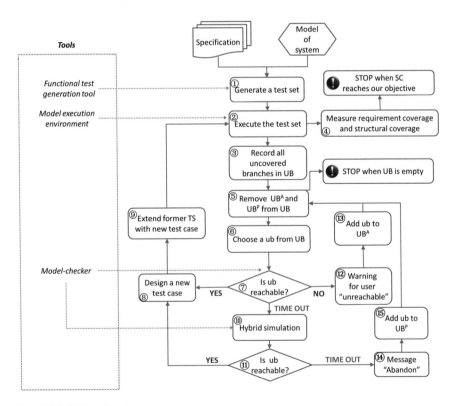

Fig. 2 Model-based testing process

about all uncovered branches in the test set is also provided. We define UB as the set of all uncovered branches after the execution of a TS (step 3). UB^A is the set of all **actual unreachable** branches and UB^P the set of all **potentially unreachable** branches, for which the method did not succeed to answer the reachability question. Initially they are both empty. Take an uncovered branch ub from UB (step 6) and apply a model-checker to perform coverability analysis [16] to check if ub is reachable (step 7):

- If ub is not reachable, send a warning message to user (step 12) and record ub as an **actual** unreachable branch (step 13): $UB_j^A = UB_{j-1}^A \cup ub$. Go to step 5 and continue the following process.
- If ub is reachable, the model-checker has probably produced a trace of inputs and outputs as a counterexample to the assertion "ub is not reachable". Use the trace built by model-checker to design a new test case (ntc) that covers this particular branch and possibly others (step 8). Complete the former TS with this new test case (step 9): $TS_i = TS_{i-1} \cup ntc$. Go to step 2 and continue the following process.

A third possibility is a "time out" situation: the model-checker takes too much time to decide if a branch is reachable. As a last solution, we propose to apply hybrid simulation as described in [18] to check the reachability of this branch (step 10). If the branch is reachable then go to step 8 and continue the following process. Hybrid simulation can also "time out" which leads to sending an "abandon" message to the user (step 14) and then record ub as a **potentially** unreachable branch (step 15): $UB_k^P = UB_{k-1}^P \cup ub$. Go to step 5 and continue the following process. Notice that in step 5, we have $UB_i = UB_i - UB_j^A - UB_k^P$.

This looping process converges when either the structural coverage reaches our objective or there are no more unexplored uncovered branches i.e. $UB_i = \emptyset$. The convergence of this process is obvious: since $TS_i \supset TS_{i-1}$, that leads to $UB_i \subset UB_{i-1}$ and $SC_i > SC_{i-1}$ because at least one more branch is covered.

It is possible that the process terminates immediately after execution of the first functional test set TS_0 if the corresponding SC_0 already reaches our objective. Otherwise, at the end of this process, if the loop at left is executed at least once, we have an improved test coverage; if the loop at right is executed at least once, i.e. $UB^A \cup UB^P \neq \emptyset$, further analysis is required since at least one branch may be suspected to be dead code or even the manifestation of a bug.

5 Application in "CONNEXION"

According to their purposes, the tools brought by partners of "CONNEXION" can be divided into three categories:

- For modelling the I&C system (build executable models corresponding to FD and RFD);

Fig. 3 The tool chain in "CONNEXION"

Activity	Methods, tools and elements		
System Modelling	High-level ES model	**Papyrus** (SysML language)	
	Process model	**Dymola** (Modelica language)	
	I&C models (several abstraction levels)	**SCADE Suite** (scade language)	
Test development	Generation of executables test cases	**MaTeLo**	Usage model
		INTERVAL	System model in xlia
	Production of a test sequence	**GATeL**	I&C model in scade
Test execution	Plateforme **ALICES** Observer **ARTiMon**		

- For developing test cases;
- For executing test cases.

They are summarized in the following Fig. 3. We present these tools very briefly. More details can be found in the corresponding references.

Papyrus [20] from CEA List: based on platform Eclipse, Papyrus offers an open source graphic editor for modelling in SysML [21]. In "CONNEXION", a system model (both Process and I&C are included) will be created in SysML with Papyrus. This model has a high abstraction level, corresponding to that of the specification of functional requirements.

INTERVAL [20] from CEA List: capable of generating executable test cases. It takes as input the system model in xLia [22], obtained by a semi-automatic transformation of system model in SysML created in Papyrus.

Dymola[1] is a commercial modelling and simulation environment based on open source Modelica[2] language. Dymola is used to modeling the process.

SCADE Suite [23] from ESTEREL Technologies has been chosen as the executable modeling tool. Based on Lustre language [24], SCADE Suite is tailored for designing critical system and its code generator is qualified under several certifications.[3]

Functional testing of the models built in SCADE suite can be as automated as possible with support of proper tools: GATeL [25] from CEA List and MaTeLo [23] from ALL4TECH have been chosen.

GATeL works on models described in Lustre language, therefore is compatible with SCADE models, and performs theoretical proof. It verifies properties that are invariant or characterizations of reachable states of the system. With a test objective expressed in an extended version of Lustre, GATeL generates a trace of inputs which drive the system to a state satisfying the test objective. In the testing process presented in this paper, GATeL will be used as the model-checker.

---

[1]http://www.3ds.com/products-services/catia/products/dymola.

[2]https://www.modelica.org/.

[3]http://www.esterel-technologies.com/products/scade-suite/.

MaTeLo is a test generation tool for statistical usage testing [26]. It works on a usage model, created manually from the specification of functional requirements. It then generates test cases and provides measurement of the requirements covered by these test cases. In the testing process presented in this paper, MaTeLo will be used as the functional test generation tool.

The test cases generated by MaTeLo do not include the expected outputs (oracle) to be compared with the actual outputs. The oracle is performed by a test observer ARTiMon [20] from CEA List. ARTiMon and the SCADE model are both integrated to a platform ALICES [27] from CORYS, which provides an environment of executing test set on the model.

6 Conclusion

The current design process of I&C systems requires a Functional Diagram and a Refined Functional Diagram before transforming to implemented programs. Verification of FD and RFD with respect to functional requirements are performed manually because FD and RFD are not executable. "CONNEXION" provides a tool chain allowing to create executable models equivalent to FD and RFD as well as to automate functional validation of these models as much as possible.

This paper presents a model-based testing process with the objective of enhancing structural coverage in functional testing. We have seen that this process also allows the detection of suspicious code branches that require analysis to determine whether they are truly unreachable or a bug is occurring in a condition guarding this branch. With support of proper tools, the functional test set can be extended by test cases derived from uncovered branches using model-checking. We consider branch coverage criterion in this article but principles and methodology can be well applied to other structural coverage criteria.

The "CONNEXION" project is recently supplied a unique set of verification tools. This enables our process and makes it applicable to I&C applications. "CONNEXION" has determined a progressively complex case study where our process will be applied. We also imagine mapping the requirement coverage of the functional test set to its structural coverage. This will support test cost reduction by reusing some test cases when some changes are being made to the system.

References

1. Enoiu, E.P., Causevic, A., Ostrand, T.J., Weyuker, E.J., Sundmark, D., Pettersson, P.: Automated test generation using model-checking: an industrial evaluation. In: ICTSS 2013
2. Collective. Cluster CONNEXION: Spécification d'un environnement de verification de la partie contrôle-commande. Livrable 2.1.2 (2014)

3. Devic, C., Morilhat, P.: CONNEXION Contrôle Commande Nucléaire Numérique pour l'Export et la rénovatION—coupler génie logiciel et ingénierie système: source d'innovations. *Génie Logiciel*, 104:2–11, mars (2013)
4. Pretschner, A., Philipps, J.: 10 Methodological Issues in Model-Based Testing. In: Broy, M. et al. (eds.) Model-Based Testing of Reactive Systems, LNCS 3472, pp. 281–291 (2005)
5. Utting, M., Pretschner, A., Legeard, B.: A taxonomy of model-based testing. Working Paper Series (2006)
6. IEC61804-2: Function blocks (FB) for process control—Part 2: Specification of FB concept, 2.0 edition (2006)
7. IEC61131-3: Programmable controllers—Part 3: Programming languages, 3.0 edition (2013)
8. IEC61513: Nuclear power plants—instrumentation and control important to safety—general requirements for systems (2011)
9. IEC60880: Nuclear power plants—instrumentation and control systems important to safety—software aspects for computer-based systems performing category A functions (2006)
10. INCOSE Systems Engineering Vision 2020. INCOSE (2007)
11. Fisher, A., Jacobson, C., Lee, E., Murray, R., Sangiovanni-Vincentelli, A., Scholte, E.: Industrial cyber-physical systems—icyphy. In: Proceedings of the Fourth International Conference on Complex Systems Design & Management, pp. 21–37 (2013)
12. Utting, M., Legeard, B.: Practical Model Based Testing: A Tools Approach. Morgan Kaufmann (2007)
13. Zhu, H., Hall, P.A., May, J.H.: Software unit test coverage and adequacy. ACM Comput. Surv. **29**(4), 366–427 (1997)
14. Piwowarski, P., Ohba, M., Caruso, J.: Coverage Measurement Experience During Function Test. In: ICSE 93 Proceedings of the 15th International Conference on Software Engineering, pp. 287–301 (1993)
15. Geist, D., Farkas, M., Landver, A., Lichtenstein, Y., Ur, S., Wolfsthal, Y.: Coverage-directed test generation using symbolic techniques. Lect. Notes Comput. Sci. **1166**, 143–158 (1996)
16. Ratzaby, G., Ur, S., Wolfsthal, Y.: Coverability Analysis Using Symbolic Model Checking, CHARME 2001. In: Lectured Notes in Computer Science, vol. 2144. Springer (2001)
17. Fantechi, A., Gnesi, S., Maggiore, A.: Enhancing Test Coverage by Back-tracking Model-checker Counterexamples. In: Electronic Notes in Theoretical Computer Science, vol. 116, pp. 199–211 (2004)
18. Memmi, G.: Integrated circuits analysis, system and method using model-checking. US Patent 7493247 (2009)
19. Rayadurgam, S., Heimdahl, M.P.: Coverage Based Test-Case Generation using Model Checkers. IEEE (2001)
20. Gallois, J., Pierron, J., Rapin, N.: Validation test production assistance. ICCSEA 2013
21. A practical guide to SysML: The Systems Modeling Language. Morgan Kaufmann/OMG Press (2011)
22. Gaudin, E.: Automatic test generation based on functional coverage. UCAAT (2014)
23. Chastrette, F., Vallee, F., Coyette, L.: Application of model-based testing to validation of new nuclear I&C architecture. ICCSEA 2013
24. Halbwachs, N., Caspi, P., Raymond, P., Pilaud, D.: The synchronous data-flow programming language LUSTRE. Proc. IEEE **79**, 1305–1320 (1991)
25. Marre, B., Arnould, A.: Test Sequences generation from LUSTRE Descriptions: GATEL. In: 15th IEEE Conference on Automated SW Engineering, pp. 47–60 (2000)
26. LeGuen, H., Thelin, T.: Practical Experiences with Statistical Usage Testing. In: Proceedings of the Eleventh Annual International Workshop on Software Technology and Engineering Practice (STEP'04)
27. Neyret, M., Dormoy, F., Blanchon, J.: Méthodologie de validation des spécification fonctionnelles du contrôle-commande—Application au cas d'étude du Système de Réfrigération intermédiaire (SRI). Génie Logiciel, hors-séries:12–25, mai (2014)

Design for Affordability from System to Programme to Portfolio Using Tradespace-Based Affordability Analysis: An Application to a Federated Satellite System Design Case Study

Marcus Shihong Wu

Abstract Affordability has become essential to the design of complex systems, programmes and portfolios due to evolving performance requirements, budget uncertainties and schedule uncertainties. This necessitates the design for affordability, which is the holistic consideration of performance, cost, schedule and other non-monetary parameters during early-phase design to perform valuable tradeoffs. By treating affordability as a high-priority ility, Multi-Attribute Tradespace Exploration and Epoch-Era Analysis can be used to synthesize a progressive tradespace-based method for finding affordable solutions and to perform affordability analysis. This method is called Tradespace-based Affordability Analysis and it can better simulate the scale, complexity and uncertainties in design, while mitigating the risk of losing valuable solutions due to oversight. Its feasibility is demonstrated through application to a case study on Federated Satellite Systems. Through a bottom-up design approach from system to programme to portfolio, an affordable solution with the desired tradeoffs across all its parameters can be obtained.

1 Introduction

With the advent of "designing for affordability as a requirement" [1], various methods have been proposed to incorporate explicit cost and schedule considerations on top of performance parameters for affordability analysis during early-phase

---

This material is an authorized derivative work of the author's graduate thesis "Design for Affordability in Defense and Aerospace Systems Using Tradespace-based Methods" (Copyright MIT, 2014) written at the Massachusetts Institute of Technology (MIT).

M.S. Wu (✉)
Defence Science and Technology Agency, 1 Depot Road, Singapore 109679, Singapore
e-mail: wshihong@dsta.gov.sg

© Springer International Publishing Switzerland 2016
M.-A. Cardin et al. (eds.), *Complex Systems Design & Management Asia*,
Advances in Intelligent Systems and Computing 426,
DOI 10.1007/978-3-319-29643-2_14

181

design of systems, programmes and portfolios. This is to minimize incidences of cost overruns and schedule delays in future. However, current methods for affordability analysis have been limited to tradeoffs between performance and total cost or performance and overall schedule, without holistic consideration of all three design drivers in tandem. Furthermore, these tradeoffs are conducted in static operating environments or in single point futures, without accounting for potential fluctuations in user needs, external constraints and operating contexts over time.

Without consideration of dynamic design elements, there is a risk of overlooking valuable design solutions that are more robust to performance, cost and schedule changes. To mitigate this risk, more emphasis can be placed on non-functional design criteria called "ilities", which are system properties that often manifest and determine value after a system is put into initial use [2]. As designing for ilities concern wider impacts on design with respect to time and stakeholders, they can better promote successful design development as compared to solely performance criteria. By defining affordability as the property of becoming or remaining feasible relative to resource needs and resource constraints over time, affordability can be treated as an ility that drives the design of more affordable yet technically sound architectures [3, 4]. With affordability as a high-priority ility, Multi-Attribute Tradespace Exploration (MATE) [5] and Epoch-Era Analysis (EEA) [6] can be used to establish a method called Tradespace-based Affordability Analysis (TBAA). Through this method, designing for affordability in systems, programmes and portfolios can ensure consistent value delivery with greater cost and schedule effectiveness.

1.1 Affordability in Systems, Programmes and Portfolios

The key principle in applying affordability as an ility is the increased emphasis on resources upfront during design. Resource needs and resource constraints, together with performance needs and performance constraints, can be identified early on wherever possible [3, 4]. A resource is defined as the aggregation of cost, schedule and other non-monetary factors necessary for architecting, development and operation. Resource needs are then the set of resource requirements elicited from stakeholders. Similarly, performance needs are the set of performance requirements. Both performance and resource needs are direct reflections of stakeholder preferences and they quantify the respective benefit and expenditures expected in a desirable design solution.

The concepts of resource and performance constraints are introduced for affordability analysis. Resource constraints are the statements of restrictions on these requirements that limit the range of feasible design solutions. Specifically, they are upper-bound restrictions imposed on the range of resources that could be made available to stakeholders. They are independent of stakeholder preferences and are often imposed to actively prevent cost and schedule overruns. Rational stakeholders and designers will prefer to have greater flexibility in their resource expenditures so as to increase their range of design solutions, but resource constraints limit that range to keep solutions within realistic time and monetary budgets. Resource constraints thus provide a more accurate and genuine depiction of stakeholders, designers and their real environments, where time and money are often tightly controlled. Examples of such restrictions include budget caps for investment cost and delivery deadlines.

Performance constraints are different as they can represent external factors that place upper or lower bounds or both on the range of performance attributes. An example of an upper-bound constraint may be reducing the firing range of a weapon for safety considerations, and an example of a lower-bound constraint may be having a minimum workspace inside a manned spacecraft for adequate maneuverability. Rational stakeholders and designers will prefer to achieve the greatest performance range possible given available technology, rather than being limited in their design choices early on by requirements not directly related to technical feasibility. Similarly, they provide a more accurate and realistic depiction of stakeholders and their real environments, with respect to what is technically possible, what is minimally required and what is allowed in design. Therefore, the identification of all constraints can prevent the initial limiting of stakeholder preferences and discarding potential designs in a myopic manner simply to meet external limitations.

Resource needs, resource constraints, performance needs and performance constraints can change over time. Design solutions become feasible if they fulfill resource needs and function within the resource constraints for a fixed context, while fulfilling performance needs and remain within performance constraints concurrently. As needs and contexts change, solutions may remain in, enter, or exit the feasible set of solutions. Affordable solutions are thus those that remain in or enter the feasible set of solutions. Therefore, the goal of affordability analysis is to identify solutions that remain feasible throughout or at least for a large part of the system, programme or portfolio lifecycle. Using this operationalization, an affordable solution will be one that is capable of satisfying changing resource requirements and resource constraints, as well as satisfying changing performance requirements and performance constraints over time.

Designing for affordability requires satisfying multiple performance, cost and schedule needs of stakeholders over time. With the recent increase in demand for emergent capabilities realized only by System of Systems (SoS), a single system alongside other homogeneous or heterogeneous systems can be part of a larger programme, which itself can be part of a larger homogenous or heterogeneous portfolio alongside multiple independent or semi-independent programmes. Due to the greater scales and complexity in programme and portfolio designs, it is in the interest of this paper to build upon earlier studies [3] and extend the conduct of affordability analysis beyond programme levels to portfolio levels.

As a result, a progressive design approach from system to programme to portfolio is required as a logical first-step solution. System-level design precedes programme-level design since the latter requires joint consideration of multiple independent or semi-independent constituent elements, which are typically the system-level design solutions and other programme-level design considerations. Finally, if the programme is part of a larger portfolio, portfolio-level design can be performed, where new or legacy assets are selected and simultaneously invested into collectively provide a set of joint capabilities [7]. The bottom-up approach thus demands the aggregation of system-level affordability for each constituent system to establish programme-level affordability and finally portfolio-level affordability, thereby providing overarching guidance to architecting emergent capabilities within realistic bounds of cost and time. Through this approach, affordable solutions for system, programme and portfolio design can be found collectively.

2 Tradespace-Based Affordability Analysis

Tradespace-based Affordability Analysis (TBAA) can be employed when applying the bottom-up approach. Leveraging the increased availability of computation power, the search for affordable solutions and affordability analysis can be conducted through a synthesis of these methods: Multi-Attribute Tradespace Exploration (MATE) [5] and Epoch-Era Analysis (EEA) [6]. MATE and EEA have been modified from their original constructions to include affordability considerations. Depending on the intent and desired rigour of analysis, both MATE and EEA, or only MATE, can be applied at each level of analysis to select affordable design solutions.

2.1 MATE for Affordability Analysis

MATE in TBAA [3, 4] is the value-driven search for affordable designs by aggregating multiple performance and resource attributes, as well as stakeholder utility for each attribute, into a single utility metric (MAU: Multi-Attribute Utility [8]) and expense metric (MAE: Multi-Attribute Expense [9]) respectively. By enabling a model-based investigation into many design alternatives, MATE allows a holistic consideration of performance utility and resource expense during early-phase design while avoiding premature fixation on point designs and narrow requirements [10]. MATE thus enables the direct exploration, comparison and analysis of many design alternatives on a single tradespace to make more informed decisions during affordability analysis.

The general procedure and data flow for MATE in TBAA is shown in Fig. 1. MATE begins with the establishment of design variables (factors within the designer's control that will drive the attributes), and epoch variables (factors that parameterize uncertain potential operating contexts, i.e. performance needs, resource needs, performance constraints, resource constraints) by stakeholders and architects of the system, programme or portfolio to be designed. To incorporate stakeholder preferences for each attribute, design-to-value mapping of performance,

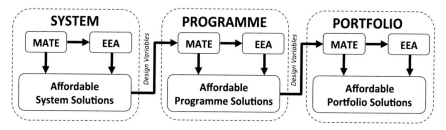

Fig. 1 Tradespace-based affordability analysis (TBAA)

as well as all resource parameters, is conducted. Applying logical assumptions and scientific principles, design and epoch variables are used to produce a tradespace model that will generate design alternatives for each epoch (period of fixed contexts, needs and constraints), and calculate the MAU and MAE to enable evaluation of the alternatives.

Tradespaces are two-dimensional plots populated by design alternatives and parameterized by their calculated MAU and MAE. With tradespaces, design alternatives can be evaluated and even compared in terms of multiple performance and resource attributes in tandem. Each performance attribute delivers a unique independent utility and can be combined with other performance attributes to produce an overall utility for a design using the MAU function. Similarly, each resource attribute has a unique independent expense and then combined in the same manner to produce an overall expense using the MAE function. MAU is a single number ranging from 0 to 1, where 0 is defined as minimally acceptable and 1 as the point where no further benefit is gained. The notion of MAE is akin to negative utility, where 0 denotes minimal dissatisfaction and 1 denotes complete dissatisfaction. Both MAU and MAE functions can be linearly weighted sums if performance and resource attributes independently contribute to the aggregate utility and expense respectively.

Performance and resource constraints are incorporated into the evaluation process after the generation of tradespaces. Upper and lower bounds on performance attributes are aggregated using the MAU function to obtain the maximum and minimum performance constraint levels respectively. Likewise, the upper bounds on resource attributes are aggregated using the MAE function to obtain the maximum resource constraint level. Shown in Fig. 2, these constraint levels are reflected as horizontal and vertical planes that segment the tradespace. If there exists an intersection between the minimum utility constraint level and a design point, the vertical planes through this design is referred to as the derived minimum expected expense constraint level. The area on the tradespace bounded by the constraint levels is the affordable solution region. An affordable design solution will then be any design point that falls within the affordable solution region. The use of constraint levels can thus help stakeholders and designers easily identify the solutions that can become unaffordable due to changes in performance or resource constraints over time.

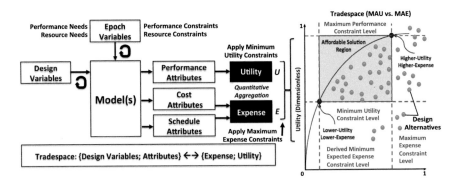

Fig. 2 Multi-attribute tradespace exploration data flow for TBAA [3, 4]

Therefore, affordable design solutions for each epoch can be obtained through MATE. However, to identify design solutions that become or remain affordable across different epochs, EEA can be applied thereafter. The affordable design solutions can then become design variables for the next level of analysis, since systems can be constituents of programmes and programmes can be constituents of portfolios.

2.2 EEA for Affordability Analysis

EEA can be used in conjunction with MATE to enable the evaluation and comparison of many different design alternatives through analysing changes in their value delivery and expense across different periods of operating contexts [3, 4]. These periods are called epochs and they are defined during MATE by the fixed set of epoch variables describing the needs and constraints in which the system, programme or portfolio operates. By analysing the affordable solution sets across different tradespaces and their corresponding epochs, design solutions that remain technically sound and affordable across all or for at least a desired number of epochs can be found. Epochs assembled into an ordered sequence form an era that can describe a potential progression of contexts over time. An era can simulate a conceivable lifecycle while multiple eras form the set of possible lifecycles that a design can undergo. EEA can thus be conducted over multiple epochs and multiple eras to evaluate system design concepts and to assess the temporal progression of performance and resource attributes for different design alternatives. Single-epoch, multi-epoch, single-era and multi-era analysis can be performed to assess the impacts of time and context on value delivery (utility and expense) under the effects of changing contexts and mission requirements.

EEA discretizes the lifecycle according to impactful changes in the operating contexts through the constructs of epochs and eras, instead of traditional system milestones. Figure 3a, b illustrate how performance-centric EEA and resource-centric

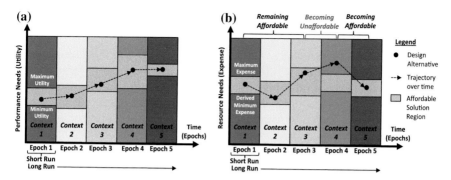

Fig. 3 **a** Original EEA for analysing performance attributes; **b** Modified EEA for analysing resource attributes [3, 4]

EEA can be conducted. In both figures, the vertical columns represent the epochs that are time-ordered to form an era, while different colours of these epochs represent changes in context. In Fig. 3a, the vertical axis measures performance needs (utility), while the minimum and maximum utility constraint levels define the affordable solution region. In Fig. 3b, the vertical axis measures resource needs (expense), while the minimum and maximum expense constraint levels further define the affordable solution region. As needs and operating contexts change, the MAU and MAE for design alternatives also change, thereby reshaping the tradespaces and affordable solution regions obtained for each epoch.

Changes in MAU and MAE for a design alternative can be represented by a trajectory in the performance space and a trajectory in the resource space respectively. The trajectory of the design alternative over time in Fig. 3a can be interpreted in the following manner: The design alternative traverses all five epochs while staying within the affordable region unique to each epoch, thereby fulfilling all performance needs and performance constraints over time. Similarly, the trajectory in Fig. 3b can be interpreted in the following manner [3]: As the design alternative traverses through the first three epochs while staying within the affordable region unique to each epoch, the system is remaining affordable. In the transition to Epoch 4, the design alternative exceeded the maximum expense constraint level, thus becoming unaffordable by the end of the epoch. Finally, the system transits back to the affordable region in Epoch 5 and is regarded as becoming affordable. With respect to EEA principles, an affordable design solution is thus one that has both performance (utility) and resource (expense) trajectories that remain within the affordable solutions defined across all epochs or for at least a desired number of epochs [4]. Therefore, MATE and EEA can be used to establish TBAA and facilitate the search for affordable design solutions.

3 Design Case Study: Federated Satellite Systems

TBAA is applied to a case study, which entailed the design and implementation of Federated Satellite Systems (FSS) [4]. The FSS comprises many heterogeneous satellite constellations or monolithic satellites in Low-Earth Orbit (LEO) working together to function as a supplier base for in-orbit data storage and processing capacity [11]. The FSS can be catered to customer spacecraft in other orbits that do not have direct access to ground stations during critical periods of time in operation and choose to transfer data via inter-satellite links (ISL) to supplier spacecraft for temporary storage or processing in situ. The FSS can thus revolutionize the design and operation of future spacecraft, which can leverage the communications and data handling capabilities of existing spacecraft in the FSS and operate without dedicated subsystems for these capabilities.

Based on a high-level assessment of its design and mission requirements, the FSS can be regarded as a homogeneous SoS that requires multiple phases of development and launching different satellite constellations to achieve its desired capability. In view of decreasing budgets and little margin for schedule overruns, it can be complex to design the FSS for affordability given the multitude of performance, cost and schedule attributes to be considered for every satellite and every constellation. Stakeholders at different levels of FSS development may also have different preferences towards performance and resource attributes depending on the operating context of their satellite constellations. Also, the development of new satellites often encounter constraints related to national security, safety, budgets and congressional decisions, which are usually considered external to the design process but can impact the selection of preferred designs.

The complexity and scale of this design problem would thus benefit from the use of TBAA in the design for affordability of the FSS. The development of the FSS could be considered as the development of a homogeneous portfolio of various satellite constellation programmes. With the application of TBAA using the bottom-up approach, preferred designs for the FSS could remain affordable at the system, programme and portfolio levels over time.

3.1 System-Level Designs and Analysis

A hypothetical development timeline for the FSS was proposed to facilitate epoch and era construction. Four operating contexts and six sets of mission needs were proposed, yielding 24 possible epochs. However, only seven of the 24 epochs were analysed. This was because the case study assumed that the development of the FSS could be conducted over seven phases, which could be described by seven sequenced epochs to form an era. This era was hypothetically imposed over the years 2016–2055. Each epoch or phase would represent 5 years, apart from the first epoch that would represent 10 years. Only one satellite constellation would be designed and launched in each epoch by a unique stakeholder with unique preferences and constraints. The FSS would thus comprise seven different satellite constellations operated by seven stakeholders, which would deliver the full operational capability over 2016–2055 (Table 1).

MATE was conducted for system-level TBAA. With knowledge of epoch variables, system design variables, system performance attributes, system expense attributes and their interactions, the system tradespace model was developed. 34560 design alternatives were generated for all 24 epochs. Stakeholder preferences

Table 1 Design Variables (DV), Performance Attributes (PA) and Resource Attributes (RA) for System-, Programme- and Portfolio-level design

System DV (Levels per DV)	System PA	System RA
• Spacecraft lifetime (2) • Initial data packet capacity (5) • Annual data packet usage rate (3) • Interface comms technology (3) • Terrestrial capacity real option (2) • Propulsion type (4) • Payload capacity (4) • Propellant mass (3) • Launch vehicle (4)	• ISL capability level • Annual available system data capacity • Annual system science value • Delta-V	• System development cost • System launch cost • System development time
Programme DV (Levels per DV)	Programme PA	Programme RA
• System design solutions (3) • Number of satellites (5) • Number of satellites in concurrent development (4) • Constellation type (3) • Choice of legacy operation (2) • Programme contract length (3)	• Annual available programme data capacity (A) • Annual programme science value (N) • Maneuverability (N)	• Total development cost (A) • Total launch cost (A) • Total labour cost (A) • Total operations cost (A) • Total retirement cost (A) • Waiting time to launch (N)
Portfolio DV (Levels per DV)	Portfolio PA	Portfolio RA
• Selected programme design solutions from each Epoch	• Annual available portfolio data capacity (A) • Annual portfolio science value (N) • Overall maneuverability (N)	• Overall development cost (A) • Overall launch cost (A) • Overall labour cost (A) • Overall operations cost (A) • Overall retirement cost (A) • Overall time to launch (N)

Aggregated attributes are denoted by (A). New attributes that represent emergent capabilities are denoted by (N)

were hypothetically established based on the different missions conducted in different contexts. Design-to-value mapping was performed and utility and expense curves were obtained for all the attributes. MAE and MAU values were then calculated for all design alternatives in every epoch, yielding 24 tradespaces. The seven tradespaces corresponding to the seven development phases were then selected and their Pareto frontiers were identified (coloured in red as shown in Epoch 1 of Fig. 4). Design alternatives not on the Pareto frontiers were removed to facilitate analysis.

Expense budget risks were also accounted for through the application of log-normal distributions with different variances to the Pareto frontier solutions. Performance and resource constraint levels were applied to determine the affordable

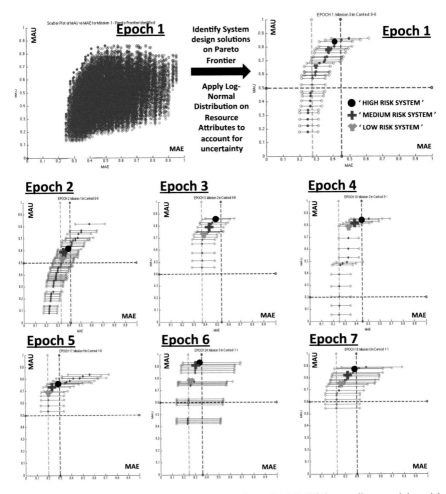

Fig. 4 Results of MATE across seven sequenced epochs [4]. High-, medium- and low-risk designs were chosen from the affordable solution region in each epoch

solution regions. No maximum performance constraints were assumed in this study. Three preferred system design solutions (High-Risk, Medium-Risk, Low-Risk) were down-selected. Their design variables and attributes were further investigated to identify commonalities that might reveal key design principles. The preferred system design solutions were the affordable system solutions and they were used as design variables for programme-level analysis. No EEA was conducted for system-level analysis.

3.2 Programme-Level Designs and Analysis

MATE was conducted for programme-level design and analysis. Programme performance and resource attributes were made up of aggregated system attributes as well as new attributes that measured emergent behaviours from operating multiple homogenous satellites in a constellation. 1080 programme design alternatives were generated. Similarly, programme design alternatives not on the Pareto frontiers were removed and lognormal distributions were applied to the remaining alternatives to simulate programme budget risk. Based on programme budget risk levels, two or three affordable programme design solutions were selected from each epoch thereafter for EEA (See Fig. 5).

Through analysing the utility and expense trajectories, programme design solutions that remained affordable from their epoch of launch and deployment were identified. Combining all identified programme solutions per epoch, 72 portfolio designs were obtained.

3.3 Portfolio-Level Designs and Analysis

Finally, portfolio-level design and analysis was conducted. MATE was performed with portfolio design variables, performance attributes and resource attributes. Similarly, these attributes were aggregates of programme attributes and represented the emergent behaviours of leveraging multiple satellite constellations in the FSS configuration to perform various science missions as well as provide onboard data processing capabilities. The portfolio design variables are the affordable programme solutions found in each epoch using EEA. No new variables or attributes were introduced at portfolio analysis. A portfolio budget was introduced as a new resource constraint, which was fixed for all portfolio designs. A tradespace with 72 portfolio designs was generated and three Pareto frontier portfolio solutions were selected for further analysis. More detailed analysis was performed to determine the resource expenditure and performance profiles of individual constellations, and ultimately the selected portfolios.

The general profiles of the three portfolio solutions were 'Mid-Utility/Mid-Expense' (Portfolio 43/72), 'High-Utility/High-Expense' (Portfolio 19)

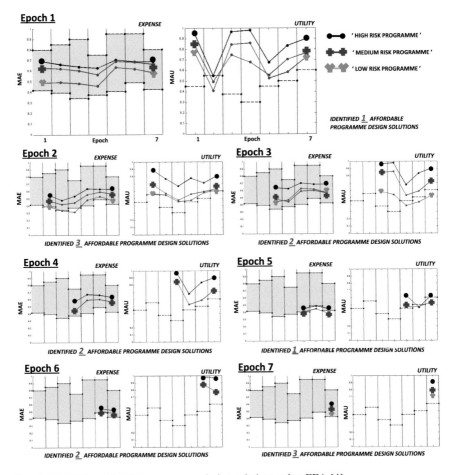

Fig. 5 Identifying affordable programme design solutions using EEA [4]

and 'Highest-Utility/Highest Expense' (Portfolio 1). The resource expenditure profile over time of each portfolio was analysed by determining the various costs and total cost committed annually to operationalizing the FSS. The performance profile over time of each portfolio was analysed by determining the overall annual data capacity and the overall science value delivered (measured in data packet volumes). Comparing and analysing the three portfolios, only Portfolio 19 was found to have not exceeded the portfolio budget at any point in time (See Fig. 6). Portfolio 1 and 43 breached the budget in one or more years during their development lifecycles. Compared to Portfolio 1, Portfolio 19 was less risky in terms of resource expenditure and offered more progressive performance levels. Compared to Portfolio 43, Portfolio 19 offered better performance during the earlier years with rising budgets and would still be able to sustain FSS operations even at its lowest data capacity. Therefore, an affordable portfolio solution (Portfolio 19), alongside affordable programme and

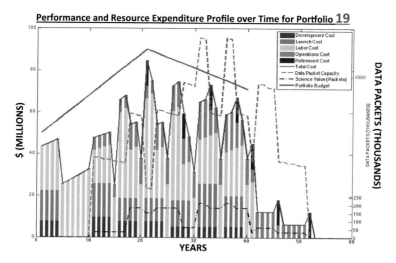

Fig. 6 Resource expenditure and performance profiles for Portfolio 19 [4]

system design solutions were collectively obtained at the end of TBAA using a bottom-up approach.

4 Discussions and Conclusion

Applying TBAA with a bottom-up approach is a logical first-step solution that can facilitate the design for affordability from system to programme to portfolio through the holistic consideration of performance needs, resource needs, performance constraints, and resource constraints over multiple epochs and multiple eras. The FSS case study demonstrated its feasibility, where a single portfolio design along with system and programme designs for the affordable development and deployment of FSS satellite constellations was obtained. For further analysis, more than three preferred designs could be picked at each level of analysis and TBAA could be conducted iteratively with affordable solution sets of different sizes until the most desirable solutions could be obtained. Therefore, TBAA could be applied to a complex SoS with homogeneous constituent systems [7] whose development roadmap resembles the bottom-up nature of the FSS. TBAA thus has immense potential in the architecting of complex systems, programmes and portfolios with greater cost and schedule effectiveness.

References

1. Carter, A.B.: Better buying power: Mandate for restoring affordability and productivity in defense spending [Memorandum] (2010). http://mil-oss.org/resources/us-dod_policy-memo_carter-memo-on-defense-spending-28-jun2010.pdf
2. de Weck, O.L., Ross, A.M., Rhodes, D.H.: Investigating relationships and semantic sets amongst system lifecycle properties (Ilities). In: 3rd International Conference on Engineering Systems. TU Delft, The Netherlands (2012)
3. Wu, M.S., Ross, A.M., Rhodes, D.H.: Design for affordability in complex systems and programs using tradespace-based affordability analysis. In: 12th Conference on Systems Engineering Research, Redondo Beach, CA (2014)
4. Wu, M.S.: Design for affordability in defense and aerospace systems using tradespace-based methods, Dual Master of Science Thesis, Aeronautics and Astronautics and Technology and Policy Program, MIT (2014)
5. Ross, A.M., Hastings, D.E., Warmkessel, J.M., Diller, N.P.: Multi-attribute tradespace exploration as a front-end for effective space system design. J. Spacecraft Rockets **41**(1), 20–28 (2004). doi:10.2514/1.9204
6. Ross, A.M., Rhodes, D.H.: Using natural value-centric time scales for conceptualizing system timelines through Epoch-Era analysis. In: INCOSE International Symposium 2008, Utrecht, The Netherlands (2008)
7. Vascik, P.D., Ross, A.M., Rhodes, D.H.: A method for exploring program and portfolio affordability tradeoffs under uncertainty using Epoch-Era analysis: a case application to carrier strike group design. In: Proceedings of the 12th Annual Acquisition Research Symposium-Acquisition Management, Monterey, CA (2015)
8. Keeney, R.L., Raiffa, H.: Decisions with Multiple Objectives-Preferences and Value Tradeoffs. Cambridge University Press, Cambridge (1993)
9. Diller, N.P.: Utilizing multiple attribute tradespace exploration with concurrent design for creating aerospace systems requirements engineering. S.M. Thesis. Massachusetts Institute of Technology, Cambridge, MA (2002)
10. Ross, A.M., Hastings, D.E.: The tradespace exploration paradigm. In: INCOSE International Symposium 2005, Rochester, NY (2005)
11. Golkar, A.: Federated satellite systems: an innovation in space systems design. In: 9th IAA Symposium on Small Satellites for Earth Observation. International Academy of Astronautics, Berlin, Germany (2013)

An Optimization Framework for Terminal Sequencing and Scheduling: The Single Runway Case

Jitamitra Desai and Rakesh Prakash

Abstract This paper addresses the static aircraft sequencing problem over the entire terminal maneuvering area (TMA) under a mixed-mode, single runway operating scenario. In contrast with existing approaches that only consider the runway as a bottleneck, our approach optimizes flight sequences and schedules by taking into account the configuration and associated constraints of the entire TMA region. This problem is formulated as a 0-1 mixed-integer linear programming problem. Efficient preprocessing and variable fixing strategies, along with several classes of valid inequalities, are derived to tighten the continuous relaxation of the problem. Computational results on illustrative examples show the overall delay in the system can be reduced by nearly a 30 % margin over the default FCFS policy and by nearly 10 % over the runway sequencing policy.

Keywords Aircraft sequencing problem · Terminal sequencing and scheduling · 0-1 mixed integer programming · Runway optimization · Air traffic management · Arrival and departure management

1 Introduction

As global business activities shift their focus towards the Asia-Pacific region, a steep increase in air traffic is expected over the next two decades, with global passenger throughput estimated to touch the nine billion mark by 2025. Air

This research was supported in part by ATMRI (NTU-CAAS) Grant No. M4061216.

J. Desai (✉) · R. Prakash
Nanyang Technological University, 50 Nanyang Avenue, Singapore 639798, Singapore
e-mail: jdesai@ntu.edu.sg

R. Prakash
e-mail: rakesh007@e.ntu.edu.sg

© Springer International Publishing Switzerland 2016
M.-A. Cardin et al. (eds.), *Complex Systems Design & Management Asia*,
Advances in Intelligent Systems and Computing 426,
DOI 10.1007/978-3-319-29643-2_15

195

transportation systems world-wide are currently operating at (or close to) capacity due to the rapid increase in air traffic demand, leading to severe congestion and delays. According to the Bureau of Transportation Statistics (BTS), in 2014, nearly one-fourth of all flights in the United States arrived late at their destination by at least 15 min, and of these late arrivals, nearly one-third were delayed due to the inability of the *Air Traffic Management* (**ATM**) system to efficiently handle the air traffic demand, and an additional one-third were caused by a concatenated effect of an arriving aircraft's delay on its next scheduled departure. Federal Aviation Administration (FAA) and Eurocontrol forecasts estimate that this imbalance in air traffic demand and existing airport capacity will continue to widen in the years ahead.

This flight delay syndrome is a very serious and widespread problem, and ever-increasing flight delays have placed a significant stress on the entire ATM system, costing airlines, passengers, and the overall economy several billions of dollars each year. In a study on the impact of delay by [1], it was estimated that, in 2007, there were direct losses of US$28.9 billion attributed to aircraft delay alone and moreover, this delay increases nonlinearly as a function of demand. The critical bottleneck within an ATM system is the capacity within a radius of about 50 nautical miles (nm) from an airport, namely the *Terminal Maneuvering Area* (**TMA**). In the United States, terminal area congestion accounted for 13 % of all delays in 2005; this number has been steadily increasing every year, and accounted for nearly 21 % of delays in 2013 (https://www.faa.gov/data_research/aviation_ data_statistics/).

There are several possible areas of improvement that can be considered for decreasing or mitigating the flight delay propagation syndrome within the TMA region. The challenge lies in simultaneously achieving safety, efficiency, and equity, which are often competing objectives [2]. Current ATM systems address the issues of passenger safety, runway efficiency, and airline equity independently, and there are few solutions that handle all of these concerns simultaneously. One way of addressing this problem is to invest heavily into airport capacity expansions, but such strategic investments require long lead times and are subject to other national and economic constraints. However, according to [3], significant delay savings can yet be achieved within existing ATM systems by optimizing critical bottleneck operations related to arrivals, departures, runways, and taxiways.

One such aspect is the joint sequencing and scheduling of arriving and departing aircraft, which is commonly referred to in the literature as the *Aircraft Sequencing Problem* (**ASP**). Specifically, in the *static version* of this problem; given a set of aircraft, along with information on the earliest/latest operation time for each aircraft (be it an arrival or a departure), and the minimum safety regulations to protect trailing aircraft from wake vortices generating by leading aircraft; the objective is to determine the optimal sequence that maximizes runway throughput or minimizes the total delay in the system, when operating under a mixed-mode of operations.

Usually, air traffic controllers follow the *First-Come First-Serve* (**FCFS**) rule for sequencing flight arrivals and departures based on their *estimated time of arrival* or *departure* (**ETA** or **ETD**) on the *runway*. But, it is well-known that such a sequencing rule is very inefficient in practice as it induces a lot of delay into the system [4]. Hence, many air traffic management advisory systems such as COMPAS [5], MAESTRO [6], and FAST [7] have been developed to assist controllers better enable procedures in sequencing aircraft by allowing *position shifting* of aircraft with respect to the FCFS sequence. More specifically, these systems implement a *constrained position shifting* (**CPS**) strategy, a concept introduced in [8], wherein an aircraft cannot be shifted by more than k positions (the so-called *maximum position shifting* (**MPS**) parameter) from its initial FCFS-based position.

Recognizing that terminal area scheduling is indeed more realistic (as compared to only runway scheduling), while also affording greater flexibility in managing aircraft, in this research effort, we optimize the sequence of arriving and departing aircraft and determine the optimal flight schedules based on the entire TMA, where the objective is to minimize the total delay in the system. Our approach is a holistic approach (addressing safety, efficiency, and equity) that presents manifold improvements as compared to earlier works, both from a modeling perspective as well as practical considerations, such as: (i) Extension of the ASP to incorporate the configuration of the TMA, thereby accounting for all safety constraints and bottlenecks in the system; (ii) Inclusion of the CPS constraint within the optimization formulation, which forms an important component in maintaining equity amongst airlines; and (iii) Our formulation also yields the advantage of determining exact solutions, as compared to fuzzy logic implementations often seen in practice, thereby attaining minimum delays and improving overall system efficiency.

The remainder of this paper is organized as follows. In Sect. 2, a brief literature review that summarizes some of the prominent works on runway and terminal area scheduling is presented. Then, in Sect. 3, we formulate the aircraft sequencing problem, defined over the entire TMA, as a 0-1 *mixed-integer linear program* (**MILP**), and several variable fixing strategies and different classes of valid inequalities that serve to improve the continuous relaxation of the 0-1 MILP are also derived. Computational results related to an illustrative example are presented in Sect. 4, and finally, Sect. 5 summarizes the contributions of this work and suggests extensions for future research.

2 Literature Review

A recent survey by [4] identifies the following essential attributes of the aircraft sequencing problem: runway operating mode, which can be segregated (only one of landings or departures) or mixed (both landings and departures); single or multiple runways; static or dynamic sequencing; and the different types of objective functions that reflect different stakeholders such as aviation authorities, airlines, airport management, etc. It is worthwhile to note that arrival management has received significantly greater attention in the literature than departure management (possibly due to the critical nature of landings), and there are just a handful of papers that address the joint sequencing problem of both arrivals and departures.

2.1 Runway Sequencing and Scheduling

Most of the existing works in the literature solve the static aircraft sequencing problem by considering only the *runway* as a bottleneck within the ATM system. In such a scenario, the required safety regulations are imposed as *time-based* separation standards between leading and trailing aircraft, which are specified in Table 1. In a pioneering work, [9] modeled the static ASP problem using *general precedence* binary variables as a 0-1 MILP considering precedence restrictions, controller workload, runway workload balancing, and number of landings in a given time period, for single and multiple runways. In [10], the authors present another 0-1 MILP formulation for the same problem using *immediate precedence* binary variables. Furthermore, taking advantage of the underlying network structure and inherent time-window restrictions present in the problem, they prescribe several preprocessing procedures, and as a result, a significant reduction in computational time, in comparison to [9], was achieved for the same data sets.

Other notable works include [11] that deals with a genetic algorithm for scheduling aircraft arrivals at Heathrow airport, and [12] developed a tabu search method for sequencing aircraft departures. [13] modeled the ASP as a shortest path network, wherein aircraft sequences are represented as nodes in the network, edge weights denote the separation times between successive aircraft, and a dynamic programming algorithm is prescribed to solve this problem. [14] proposed genetic algorithms to efficiently schedule aircraft landings and reported empirical results for different scenarios. [15] modeled the static ASP problem as a traveling salesman problem (TSP); [16] presented a MILP formulation for the single runway aircraft landing problem; and [17] modified the earlier work of [15] by taking earliest/latest times into consideration. Finally, [18] studied an aircraft landing problem over a single runway with holding patterns, where an aircraft is assigned a set of disjoint time-windows and the objective function is to minimize the maximum time between consecutive landings.

Table 1 Time-based separation standards enforced at the runway for various cases of leading/following aircraft

Leading/following	Departure → Departure			Departure → Arrival		
	Heavy	Large	Small	Heavy	Large	Small
Heavy	60	90	120	50	53	65
Large	60	60	90	50	53	65
Small	60	60	60	50	53	65
Leading/following	Arrival → Departure			Arrival → Arrival		
	Heavy	Large	Small	Heavy	Large	Small
Heavy	40	40	40	99	133	196
Large	35	35	35	74	107	131
Small	30	30	30	74	80	98

2.2 Terminal Sequencing and Scheduling

To the best of our knowledge, only the work of [19] has considered solving the aircraft sequencing problem over the entire TMA region. They formulate this problem as a *job-shop scheduling model* with sequence dependent set-up times and release dates to schedule both arrivals and departures. More specifically, in this job-shop model, aircraft at different holding points near the entry fixes are treated as *jobs*, which are waiting to be scheduled and the trajectory segments that lie along a route from an entry gate to the runway, are modeled as a series of *machines*. Once a 'job' is released from its holding stack, it flows freely from one 'machine' to the next until it reaches the runway, which becomes the final machine for arriving aircraft and the initial machine for departing aircraft. However, they do not consider the use of time-windows for scheduling arrivals or departures, which greatly limits the practical applicability of their model. More importantly though, their computational results reveal that a bottleneck can occur anywhere in the TMA and not only on the runway threshold.

During the development of the *Final Approach Spacing Tool* (**FAST**) software, [7] found that merge points within the TMA cannot be ignored from a scheduling perspective because optimal sequencing based solely on runway constraints results in schedules that require heavy conflict detection and upstream resolution tasks. Therefore, the FAST algorithm begins by generating feasible sequences and schedules, and subsequently resolves conflicts that might arise by manipulating the trajectories of the pair of aircraft in conflict. When resolving such conflicts, the algorithm in FAST adds delays to the trailing aircraft and, in an indirect manner, adds to controller workload by issuing vectoring instructions.

Motivated by these earlier works on aircraft sequencing problems, in this paper, we present a mathematical programming framework for solving the joint arrival-departure sequencing and scheduling problem over the entire TMA region, where the objective is to minimize the total delay in the system.

3 Modeling and Analysis

The *Terminal Manoeuvring Area* (**TMA**) is a circular region having a 50 nm radius centered at the airport. At the boundary of the TMA, there are various *way-points* or *fixes*, labeled as entry (or meter) gates, through which traffic enters the TMA. The traffic from the enroute airspace merges at each entry gate and flows into the TMA as *one stream* from each gate. Given a stream, there is a pre-specified route leading to the runway, where a route can be best defined as a series of way-points, and where the sections of a route between two consecutive way-points is said to form a *trajectory segment*. (Multiple routes may share one or more of such trajectory segments.) An incoming aircraft from the enroute airspace either enters into the holding area near the entry gates or proceeds directly along a pre-specified route to its allocated runway. Traffic from different streams merges together at common way-points (merge points) and subsequently flows along the same route. Finally, all traffic designated for the same runway merges into a single stream at the *final approach fix* (**FAF**) and descends into the runway threshold. (This operation is normally performed at a distance of 5 nm from the runway.) Similarly, for departures, aircraft flow together towards the *initial fix* near the runway, and then diverge along different trajectory segments at way-points, finally exiting the TMA from the exit gates.

Once an approach procedure is initiated at the entry gates, a switch in position between aircraft on the same route is not permitted, and furthermore, during this entire approach operation, safety separation standards must be maintained between consecutive and non-consecutive pairs of aircraft. In contrast to the time-based separation standards enforced on the runway (see Sect. 2), the FAA-specified longitudinal safety separation requirements in the TMA region for both arrivals and departures are *distance-based*, which are displayed in Table 2.

As shown in Fig. 1, the TMA region has been represented as a *graph*, where the waypoints are shown as *nodes* or (*vertices*) and the trajectory segments between two consecutive waypoints as *edges*. Arrival traffic enters into the TMA through various entry gates and departure traffic enters into the runway through a holding

Table 2 Distance-based separation standards (in miles) enforced within the TMA for both arrivals and departures

Leading/Following	Heavy	Large	Small
Heavy	4	5	6
Large	3	3	4
Small	3	3	3

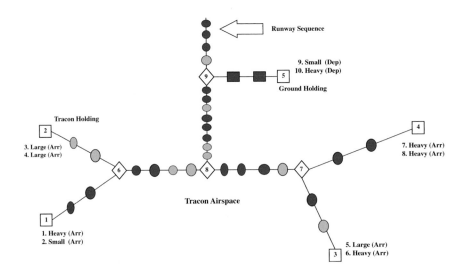

Fig. 1 A graphical representation of a TMA region with four entry fixes, four merge points, and a single runway

area, which is adjacent to the runway. It has been well-established that absorbing anticipated delays at higher altitudes, is always better from fuel consumption, safety, and controller workload viewpoints [20]. Hence, in our model representation, we work under the assumption that delay or earliness is induced in an aircraft before entering the TMA, either by adjusting its velocity or by holding at the entry gates. (For departure traffic, delay is induced on the ground.) As aforementioned, when compiling the final runway sequence, the nominal order of aircraft in the same stream is always preserved, thereby ensuring that overtaking is avoided and a runway closest to the entry fix of the aircraft is allocated.

The aircraft sequencing problem defined over the TMA can now be formulated, where the definitions of the index sets, parameters, and decision variables are given as follows.

Description of Index Sets and Parameters

- $n \in \mathcal{N} \equiv \{1, 2, \ldots, N\}$: Set of all nodes in the graph representing entry gates, merge points, departure holding area, and the runway
- $g \in \mathcal{G}$: Set of all the nodes in the graph representing entry gates, where $\mathcal{G} \subseteq \mathcal{N}$
- $(p, q) \equiv e \in \mathcal{E}$: Set of all the edges (trajectory segments) in the graph, where $p \in \mathcal{N}, q \in \mathcal{N}$
- r: Vertex representing runway
- $f \in \mathcal{F}$: Set of all arriving and departing flights
- v_i^e: Velocity of any flight i in trajectory segment $e \in \mathcal{E}$
- $\mathcal{F}^n$: Set of flights passing through vertex n, where $\mathcal{F}^n \subseteq \mathcal{F}$
- E_i^n: Earliest time of arrival (departure) of aircraft i at (from) vertex n
- T_i^n: Target time of arrival (departure) of aircraft i at (from) vertex n
- L_i^n: Latest arrival (departure) of any aircraft i at (from) vertex n
- Δt_{ij}: Minimum safety separation (in seconds) at runway threshold if flight i is ahead of flight j (see Table 1)
- Δs_{ij}: Minimum safety separation (in miles), if flight i is ahead of flight j (see Table 2)
- ROT_i: Runway occupancy time of aircraft i
- seq_i: Position of flight i based on the FCFS sequence
- k: Specified maximum position shifting (MPS) parameter

Decision Variables

- $x_{ij}^n = \begin{cases} 1, & \text{if flight } i \text{ is ahead of flight } j \text{ in schedule at vertex } n \\ 0, & \text{otherwise} \end{cases}$
- t_i^n = Scheduled time of arrival (departure) of flight i at (from) vertex n

$$\textbf{ASP-TMA} \quad \text{Minimize} \sum_{f \in \mathcal{F}} |t_f^r - T_f^r|. \tag{1a}$$

$$\text{subject to} \quad x_{ij}^n + x_{ji}^n = 1, \quad \forall i < j, (i,j) \in \mathcal{F}^n, \forall n \in \mathcal{N}. \tag{1b}$$

$$t_i^q = t_i^p + T_i^q - T_i^p, \quad \forall i \in \mathcal{F}^p \cap \mathcal{F}^q, \forall (p,q) \in \mathcal{E}. \tag{1c}$$

$$t_j^p \geq t_i^p + (\Delta s_{ij}/v_i^e) - M(1 - x_{ij}^p), \quad \forall i,j \in \mathcal{F}^p. \tag{1d}$$

$$t_j^q \geq t_i^q + (\Delta s_{ij}/v_j^e) - M(1 - x_{ij}^q), \quad \forall i,j \in \mathcal{F}^q, \ \forall (p,q) \in \mathcal{E}. \tag{1e}$$

$$t_j^r \geq \max\{\Delta t_{ij}, ROT_i\} + t_i^r - M(1 - x_{ij}^r), \quad \forall i,j \in \mathcal{F}. \tag{1f}$$

$$-k \leq (F - \sum_{j \in \mathcal{F}, i \neq j} (x_{ij}^r)) - seq_i \leq k, \quad \forall i \in \mathcal{F}. \tag{1g}$$

$$x_{ij}^q = x_{ij}^p \quad \forall i < j, (i,j) \in \mathcal{F}^p, \ \forall (p,q) \in \mathcal{E}. \tag{1h}$$

$$E_n^i \leq t_n^i \leq L_n^i \quad \forall i \in \mathcal{F}^n, \ \forall n \in \mathcal{N}. \tag{1i}$$

$$x_{ij}^n \in \{0,1\}, \quad t_i^n \geq 0 \quad \forall i,j \in \mathcal{F}^n, \ \forall n \in \mathcal{N}. \tag{1j}$$

In the above formulation, the objective function (1a) seeks to minimize the total delay; constraint (1b) enforces the order precedence between flights i and j at every vertex; constraint (1c) computes the scheduled time of arrival (or departure) of flight i arriving at node q from node p, where $(p,q) \in \mathcal{E}$, based on the condition of free flow of flights once they enter the TMA; constraint (1d) ensures the longitudinal separation requirement (in miles) between flights departing from each $n \in \mathcal{N} \backslash \{r\}$; constraint (1e) ensures the longitudinal separation requirement (in miles) between flights arriving at each $n \in \mathcal{N} \backslash \{r\}$; constraint (1f) ensures the time-based separation requirement between flights at the runway node r; constraint (1g) imposes the CPS constraint that an aircraft cannot be shifted by more than k positions from its initial (FCFS) position, where $F \equiv |\mathcal{F}|$; constraint (1h) avoids overtaking by maintaining the precedence relationship between two aircraft at succeeding way points; constraint (1i) maintains the scheduled time of arrival (departure) at each vertex to be between the earliest and latest times for each aircraft; and finally constraint (1j) imposes the binary and non-negativity restrictions on the x—and t— variables, respectively.

The ASP-TMA formulation described above can be further enhanced by pre-fixing some of the variables and by addition of valid inequalities that serve to tighten the underlying linear programming representation of the problem. In the following discussion, we derive several model enhancing features based on time-window restrictions, maintenance of nominal order of aircraft along the same stream, and relative position of an aircraft within the sequence. These strategies potentially help in obtaining a partial convex hull representation, and were therefore found to be very effective in our computations.

Variable Fixing Strategies

Proposition 1 *At any vertex n, consider a pair of aircraft $i, j \in \mathcal{F}^n$. If the FCFS sequence positions of i and j satisfy the condition: $seq(j) - seq(i) \geq 2k$, where k is the MPS parameter, then we can fix $x_{ij}^n = 1$.* □

Proposition 2 *For a pair of aircraft $(i, j) \in \mathcal{F}^n$, if $E_j^n > L_i^n$, or more generally, if $E_j^n + sep_{ji}^n > L_i^n$, we set $x_{ij}^n = 1$, where sep_{ji}^n denotes the to-be-maintained separation between a leading aircraft j and a following aircraft i.* □

Proposition 3 *At any entry waypoint g, consider a pair of aircraft $i, j \in \mathcal{F}^g$ that belong to the same category. Suppose that the time-window restrictions satisfy the following conditions:*

(i) $E_i^g \leq E_j^g$; (ii) $T_i^g \leq T_j^g$ and (iii) $L_i^g \leq L_j^g$. Then, we can set $x_{ij}^g = 1$. (It should be noted that this prefixing routine cannot be applied at merge points because flights originating at different entry fixes are in negotiation at each intermediate node.) □

Valid Inequalities

In order to improve the quality of the lower bound, decrease the required computational effort, and reduce the search space, we derive a set of valid inequalities, that are given below.

- Let $i, j \in \mathcal{F}^n$ be a pair of aircraft such that, $(L_j^n - E_i^n) \equiv \max\left\{t_j^n - t_i^n : \text{constraint (1i)}\right\} > 0$. If $t_j^n - t_i^n > 0$, we can enforce $x_{ij}^n = 1$ by using the following valid inequality:

$$x_{ij}^n \geq (t_j^n - t_i^n)/(L_j^n - E_i^n) \quad \forall i < j, i, j \in \mathcal{F}^n, L_j^n > E_i^n, n \in \mathcal{N}. \quad (2a)$$

- At any vertex n, clearly,

$$\sum_{i \in \mathcal{F}^n} \sum_{i \neq j, j \in \mathcal{F}^n} (x_{ij}^n) = F^n(F^n - 1)/2, \quad \forall n \in \mathcal{N}, \text{ where } F^n \equiv |\mathcal{F}^n|. \quad (2b)$$

- At any vertex n, given a 3-tuple of aircraft $(i, j, k) \in \mathcal{F}^n$, we can impose the transitive constraints that if aircraft i is ahead of aircraft j, which in turn is ahead of aircraft k, than it implies that i is ahead of k. The following valid inequalities are satisfied by any of the six possible (i, j, k) arrangements of planes.

$$x_{ik}^n \geq x_{ij}^n + x_{jk}^n - 1 . \tag{2c}$$

$$x_{ik}^n \leq x_{ij}^n + x_{jk}^n \quad \forall i > j, j > k, (i,j,k) \in \mathcal{F}^n, n \in \mathcal{N} . \tag{2d}$$

We refer to the ASP-TMA formulation, given by (1a)–(1j), reinforced with all these variable fixing strategies and valid inequalities (2a)–(2d) as RASP-TMA.

4   Computational Results

For the sample TMA configuration shown in Fig. 1, we begin by considering a set of 10 aircraft, belonging to different weight categories, comprising eight arrivals and two departures. The earliest time of arrival (or departure) for all aircraft has been set as 100 s before their respective target time and the latest time is taken to be 1 h beyond the target time. The velocity of heavy, large and small categories of aircraft is given by 260, 160, and 140 nm/h, respectively, and it is assumed to be constant throughout the TMA. Arriving flights either stack up at entry gates (labeled as nodes $\{1, 2, 3, 4\}$), or enter the TMA, merging with other streams at merge points (labeled as nodes $\{6, 7, 8\}$), before finally merging with scheduled departures (at node labelled $\{9\}$), resulting in the final runway sequence.

We tested the relative effectiveness of three different sequencing models: FCFS; optimal runway sequencing (obtained by ignoring the TMA configuration constraints and enforcing only the runway separation standard in ASP-TMA); and the proposed ASP-TMA formulation. Figures 2 and 3 display the optimal sequences and associated delays obtained, using the aforementioned sequencing algorithms, corresponding to the MPS parameter $k = 1$ and $k = 3$ scenarios. As seen in Fig. 2, allowing an aircraft to shift its position by even one unit, with respect to its FCFS sequence-based position, can result in a significant reduction in total delay, and the accrued delay savings increase with an increase in the value of k (see Fig. 3). This result is further validated in Fig. 4, which plots the percentage delay savings obtained by the ASP-TMA model over the FCFS sequence as a function of k, for varying number of aircraft.

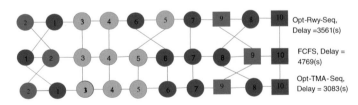

Fig. 2 Optimal sequences obtained by the three sequencing models for the case of $k = 1$

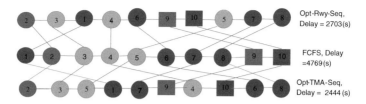

Fig. 3 Optimal sequences obtained by the three sequencing models for the case of $k = 3$

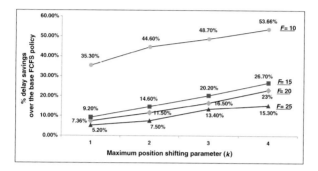

Fig. 4 Percentage delay savings as a function of MPS parameter k for varying numbers of aircraft

Having ascertained the efficacy of the proposed approach, we now examine the strength of the variable fixing strategies and derived valid inequalities by comparing the performance of ASP-TMA and the reinforced RASP-TMA models. Table 3 records various computational parameters, namely the CPU time (in seconds) taken to determine the optimal solution, branch-and-bound nodes enumerated and the LP-IP gap% at the root node, as a function of problem size and the MPS parameter k for these two models. As seen from the results, employing these preprocessing steps, leads to a significant reduction across all measured parameters with the LP-IP gap% having decreased on average to 81.6 from 100 %, when using the enhanced RASP-TMA formulation. Moreover, all of the larger test instances that could not be solved (when using a time limit of 3600 s) using ASP-TMA, were easily solved when reinforced with these additional strategies. Furthermore, it is worthwhile to note that the computational effort required increases significantly with an increase in k, even within data sets having the same number of aircraft. As a practical recommendation, based on our computational experience, we recommend using the RASP-TMA model with no more than a maximum position shifting parameter value of $k = 2$.

Table 3 Computational comparisons between ASP-TMA and RASP-TMA

		ASP-TMA			RASP-TMA		
No. of aircraft	k	CPU time (s)	Nodes	LP-IP gap (%)	CPU time (s)	Nodes	LP-IP gap (%)
10	1	0.07	615	100	0.02	15	55.9
	2	0.34	2037	100	0.06	174	88
	3	0.53	5153	100	0.14	1617	90.3
	4	1.06	13986	100	0.36	3306	92.3
15	1	1.04	2724	100	0.03	126	56.9
	2	6.74	52687	100	0.2	3852	85.6
	3	106.26	1304660	100	1.06	10780	92.3
	4	407.55	6739628	100	3.04	34011	92.3
20	1	4.30	5833	100	0.16	92	55.5
	2	421.74	2045306	100	1.53	10494	87.7
	3	>3600	–	100	8.43	42121	92.7
	4	>3600	–	100	23.39	115008	93
25	1	38.19	29214	100	0.16	1342	55.3
	2	>3600	–	100	1.96	23853	89
	3	>3600	–	100	32.95	41976	95
	4	>3600	–	100	141.87	861072	95.38

5   Conclusion

In this paper, we posed and solved the aircraft sequencing problem over the entire TMA region taking into account several realistic constraints such as longitudinal separations at the runway and in the TMA airspace, CPS constraints, and precedence maintenance requirements, with the objective of minimizing the total delay in the system. This complex problem was formulated as a 0-1 MILP, and several variable fixing strategies and valid inequalities are derived to improve the computational efficiency of the model. Our results indicate that significant delay savings can be achieved over the base FCFS policy schedule and over the runway optimized sequence. Our model is generic enough to be applied to different types of TMA configurations and can handle any mixture of traffic for the single runway case under segregated or mixed-modes of operations.

There are several potential areas of improvement that can be considered for future research, both from modeling and algorithmic perspectives. There are several classes of convex hull defining valid inequalities that can be gainfully employed to improve the model representation. Moreover, our formulation appears amenable to being solved via a Lagrangian dual approach. Furthermore, we can also extend this model to the multiple runway scenario (which is the case in most major airports). We are currently working on all of these improvements, and finally, a realistic computational case study at Changi airport is also being considered.

References

1. Ball, M., Barnhart, C., Dresner, M., Hansen, M., Neels, K., Odoni, A., Peterson, E., Sherry, L., Trani, A., Zou, B.: A comprehensive assessment of the costs and impacts of ight delay in the united states. NEXTOR (2010)
2. Bianco, L., DellOlmo, P., Giordani, S.: Scheduling models and algorithms for TMA traffic management. Modeling and Simulation in Air Traffic Management, Springer, pp. 139–167 (1997)
3. Soomer, M., Frank, G.: Scheduling aircraft landings using airlines preferences. Eur. J. Oper. Res. **190**, 277–291 (2008)
4. Bennell, J.A., Mesgarpour, M., Potts, C.N.: Airport runway scheduling. 4 OR: Q. J. Oper. Res. **9**(2), 115–138 (2011)
5. Volckers, U.: A comprehensive assessment of the costs and impacts of flight delay in the United States. In: Proceedings of the 1990 American Control Conference (1990)
6. Garcia, J.: Maestro-Metering and spacing tool. Proceedings of the 1990 American Control Conference (1990)
7. Davis, T.J., Krezowoski, K.J., Bergh, C.: The final approach spacing tool. In: 13th IFAC Symposium on Automatic Control in Aerospace (1994)
8. Dear, R.G.: The dynamic scheduling of aircraft in the near terminal area. Technical report, R-76(9) (1976)
9. Beasley, J.E., Krishnamoorthy, M., Sharaiha, Y.M., Abramson, D.: Scheduling aircraft landings: the static case. Trans. Sci. **37**(4), 180–197 (2000)
10. Ghoniem, A., Sherali, H., Baik, H., Trani, A.: A combined arrival-departure aircraft sequencing problem. INFORMS J. Comput. **26**(3), 514–530 (2014)
11. Beasley, J., Sonander, J., Havelock, P.: Scheduling aircraft landings at London heathrow using a population heuristic. J. Oper. Res. Soc. **52**(5), 483–493 (2001)
12. Atkin, J., Burke, E., Greenwood, J., Reeson, D.: Hybrid metaheuristics to aid runway scheduling at London Heathrow. Trans. Sci. **41**(1), 90–106 (2007)
13. Balakrishnan, H., Chandran,B.: Algorithms for scheduling runway operations under constrained position shifting. Oper. Res. **58**(6), 1650–1665 (2010)
14. Hansen, J.: Genetic search methods in air traffic control. Comput. Oper. Res. **31**(3), 445–459 (2004)
15. Psaraftis, H.N.: A dynamic programming approach to the aircraft sequencing problem. MIT Flight Transportation Laboratory Report R76–9 (1976)
16. Abela, J., Abramson, D., Krishnamoorthy, M.: Computing optimal schedules for landing aircraft. In: Proceedings of the 12th National ASOR Conference (1993)
17. Venkatakrishnan, S., Barnett, A., Odoni, A.R.: Landings at Logan airport- Describing and increasing airport capacity. Trans. Sci. **27**, 211–227 (1993)
18. Artiouchine, K., Baptiste, P., Durr, C.: Runway sequencing with holding patterns. Eur. J. Oper. Res. **189**, 1254–1266 (2008)
19. Bianco, L., DellOlmo, P., Giordani, S.: Scheduling models for air traffic control in terminal areas. J. Sched. **9**(3), 223–253 (2006)
20. Erzberger, H.: Design principles and algorithms for automated air traffic management. Knowledge Based Functions in Aerospace Systems, AGARD Lecture Series no. 200 (1995)

Holistic Integrated Decisions Trade-Off Baseline for an Optimal Virtual IP Multimedia Subsystem (IMS) Architecture

Arevik Gevorgyan, Daniel Krob and Peter Spencer

Abstract Network Functions Virtualization (NFV) reveals the challenge of transformation from traditional monolithic to optimal virtual architectures, while meeting the standards-driven and diverse stakeholders functional, performance constraints and conflicting objectives: i.e. maximizing flexibility (scaling, growth), minimizing expense in terms of financial and human effort and intervention (CapEx, OpEx), etc. We case study the IP Multimedia Subsystem, as being the most complex and important NFV instance. Our approach, specifying a customizable baseline to arbitrate and to subsequently optimize between the NFV strategic objectives and enablers, defines an optimal functional architecture adapted to the context at any level and scope: from a specific virtual network function resources (virtual machines) till infrastructural (multiple virtual IMS systems) requirements. To trade-off for the optimal functional options, we explore the multi-objective optimization method, incorporated with the data gathered from an adapted Model-Based Architectural Framework, PESTEL and FURPSE analyses. We further discuss our vision: self-regulating system and the related constraints.

Keywords Network functions virtualization · IP multimedia subsystem · Optimal functional architecture · Decisions trade-off · Multi-objective optimization · Model-based architectural framework

A. Gevorgyan (✉) · D. Krob
Laboratoire d'Informatique (LIX), École Polytechnique, Route de Saclay,
91128 Palaiseau, France
e-mail: gevorgyan@lix.polytechnique.fr; arevik.gevorgyan@alcatel-lucent.com

D. Krob
e-mail: dk@lix.polytechnique.fr

A. Gevorgyan · P. Spencer
Strategy—IP Platforms, Alcatel-Lucent International, 148/152 Route de la Reine,
92100 Boulogne-Billancourt, France
e-mail: peter.spencer@alcatel-lucent.com

© Springer International Publishing Switzerland 2016
M.-A. Cardin et al. (eds.), *Complex Systems Design & Management Asia*,
Advances in Intelligent Systems and Computing 426,
DOI 10.1007/978-3-319-29643-2_16

209

1   Introduction

Network Functions Virtualization (NFV) is the greatest transformation in Telecommunications industry. It transcends the architectural, service/value, economic paradigms, while "enabling network access to a scalable and elastic pool of sharable physical or virtual resources with self-service provisioning and administration on-demand" [ISO-IEC_17788]. NFV redefines the ways of thinking and engineering required to build the networks.

Many factors can potentially contribute to the success or failure of the NFV transformation, which depends not only on network solution providers, but other direct and indirect stakeholders. Diverse stakeholders (even within one organization) envisage different NFV solutions and benefits within different timeframes. Telecom Operators, for instance, within the upcoming 5 years, are likely going to converge around the major axes:

- Maximizing Capacity and Flexibility (Scaling, Growth)
- Maximizing Capabilities and Top Line Growth
- Minimizing Capital Expenditure (CapEx) Reduction
- Minimizing Operational Expenses (OpEx) Reduction
- Maximizing Organizational Transformation

The researched unified solution shall define the dynamics of functional organization, seeking to satisfy the stakeholders diverse, mostly conflicting objectives. From the other hand, satisfaction of those objectives signifies a multi-criteria trade-off for subsequent optimizations of their NFV enablers: maximization of resources orchestration (elasticity and placement), maximization of automation (self-regulation), maximization of dynamic resiliency (scaling and growth), maximization of end-to-end efficiency (capacity and performance), minimization of customizations (service efforts) etc.

And a network of constraints of a very different nature must be analyzed to attain an architecture best adapted to the context, starting, first of all, from standards-driven functional and performance constraints.

We case study the IP Multimedia Subsystem (IMS), as the most complex and important NFV instance for Alcatel-Lucent. It is essential for communication services across networks and the only for Voice over 4G/5G. As explained in [1], achievement of an optimal IMS architecture is difficult due to the principal axes of complexities:

- IMS system (*network elements specificities, interoperability issues, etc.*),
- Organization (*traditional silo decompositions, distributed geography, etc.*),
- Direct market (*heterogeneous technologies, different maturity levels, etc.*),
- Overall evolving systemic environment, comprising diverse stakeholders and Eco-systems (*regulatory/standardization bodies, users populations, etc.*).

In order to constitute a holistic integrated functional design approach, we hence reflect upon a set of interrelated metaphors for a frame-worked analysis and optimization of the multi-facet problem. As explained above, a chain of subsequent optimizations, representing different abstraction layers, are to be resulted from the trade-off analysis outcomes. Hence, our approach shall define a customizable policy baseline to satisfy the NFV strategic objectives through arbitration and optimization of associated NFV enablers and resources.

Throughout all decision levels and spectrums, we refer to an adapted Model-Based Architectural Framework (AF) (inspired by SAGACE) to analyze and model the operational [1], functional and structural views of the virtual IMS. We incorporate it with PESTEL (*Political, Economic, Social, Technological, Ecological, and Legal*) [INCOSE] and FURPSE (*Functionality, Usability, Reliability, Performance, System Maintainability, and Evolution*) [ISO/IEC 9126] analysis frames to model the IMS environment, direct market specifics, comprising the procedure of stakeholders needs analysis.

We use AF also as a reference guideline to enhance cooperation and coordination in such a complex multi-disciplinary project. Data gathered from the structured and interrelated views of AF serves for the further re-use, justification and optimization of decisions. To attain a consensus between the multi-disciplinary needs and decisions, we hence explore optimization methods that found practical application in Systems Engineering (SE). We opt for the Multi-Objective Optimization (MOO) method to trade off in accordance to the evolving market, technological and other environmental criteria and to respectively constitute the functional choices. We further illustrate our approach trough a brief example of the virtual IMS lifecycle "Operate" phase, to better highlight the NFV innovation features.

The motivation of our approach is explained in Sect. 2 and the analysis procedure represented through a brief example in Sect. 3. In Sect. 4 we discuss our future vision and associated theoretical prospects.

2 Motivation of Our Approach

Throughout the system's lifecycle, Architectural framework (AF) provides the complete, organized information sets about the system and its stakeholders, at the same time helping to improve the multi-disciplinary project collaboration.

In our study, we refer to Penalva's SAGACE architectural framework [2], which allows an iterative and incremental procedure for a complete design and proposes a *modeling approach, matrix of nine points of view (operational, functional and structural views, all refined by three perspectives of time) and a graphical modeling language.*

Instead of time dimension, we refine the views by behavioral perspective, in order to use the SysML modeling language, as explained in [3, 4]. As explained by [5], the standardized SysML and UML modeling languages help for more effective collaborations. Implementation of each Architecture diagramme in SysML is explained by [3, 4] (see Fig. 1):

Our case study on IP Multimedia Subsystem illustrates the typical industrial problems and challenges, hence the utility of the proposed approach. Main difficulty is in the definition of adequate objectives, constraints, variables, their interrelations, the right priorities and criteria to evaluate those options (Table 1).

As industrial practice shows, when the system is complex and of large-scale, the problem space has a tendency to evolve even more and it becomes difficult to resolve. Moreover, in uncertainty situations, it is even more difficult to make a correct trade-off. In practice, optimization techniques are more efficient when the problem is relatively small and has good mathematical properties.

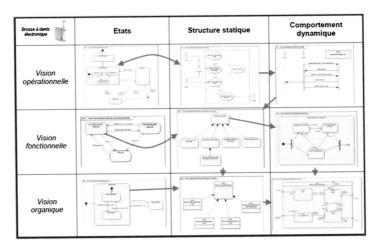

Fig. 1 SysML diagrammes incorporated in an Architectural Framework (Refs. [3, 4])

Table 1 Macro needs

N1	Operators want a system that will support significantly higher traffic loads
N2	Operators want a robust system
N3	Operators want assurance for the maintenance and support
N4	Operators want capabilities to easily deploy/support new applications/services
N6	Operators want significant savings in CAPEX/OPEX
N7	Operators want operational easiness: i.e. to drastically reduce time to market
N8	Providers want to follow existing standards and regulations
...	....

In our study, we are interested to explore the Multi-Objective Optimization (MOO) method, which is known in decision making practices. Four major categories of MOO methods are defined by Marler and Arora (2004):

- methods with a priori articulation of preferences
- methods with a posteriori articulation of preferences
- methods with no articulation of preferences
- genetic algorithms

The MOO problem is reproduced and formulated by the following equation (Eq. 1):

$$\min \mathbf{J}(\mathbf{x}, \mathbf{p}) \qquad \text{where } \mathbf{J} = [J_1(\mathbf{x}) \ \dots \ J_2(\mathbf{x})]^T$$
$$\text{s.t. } \mathbf{g}(\mathbf{x}, \mathbf{p}) \leq 0$$
$$\mathbf{h}(\mathbf{x}, \mathbf{p}) = 0 \qquad \mathbf{x} = [x_1 \ \dots \ x_l \ \dots \ x_n]^T$$
$$x_{i,LB} \leq x_i \leq x_{i,UB} \ (i = 1, \dots, n) \qquad \mathbf{g} = [g_1(\mathbf{x}) \dots g_{m_1}(\mathbf{x})]^T$$
$$\mathbf{x} \in \mathbf{S} \qquad \mathbf{h} = [h_1(\mathbf{x}) \dots g_{m_2}(\mathbf{x})]^T$$

*Here, **J** is a column vector of **z** objectives, whereby **Ji** ÎR. The individual objectives are dependent on a vector **x** of n design variables as well a vector of fixed parameters, **p**. The individual design variables are assumed continuous and can be changed independently by a designer within upper and lower bounds, **xUB** and **xLB**, respectively. In order for a particular design **x** to be in the feasible domain **S**, both a vector **g** of **m1** inequality constraints, and a vector **h** of **m2** equality constraints have to be satisfied. The problem is to minimize—simultaneously—all the elements of the objective vector [de Weck 2006].*

In a MOO problem, there is no single global solution, and it is often necessary to determine a set of points that all fit a predetermined definition for an optimum.

Pareto optimality is predominant in defining the optimal point (Marler and Arora 2004). Kim and de Weck (2005a, b), Smaling and de Weck (2004), Smaling (2005) illustrate Pareto modeling examples in systems engineering.

Hence, we shall model our decision problem as a Pareto model, without putting any preferences on the objectives of the set, in order to find the best suitable values of dimensional parameters.

Important is to note that MOO problems request clear definition and distinction between the measures of Objectives:

- *Measures of Effectiveness* (*MOE*) estimate achievement of the system's mission or operational objectives (under a specified set of conditions) within the corresponding operational environment. They are derived during the Operational Analysis phase [INCOSE 2010].
- *Measures of Performance* (*MOPs*) should be derived from or provide insight upon MOEs or User needs [INCOSE 2010]. MOPs are the key performance characteristics the system must have. They assess whether the system satisfies design or performance (i.e. technical, economic, etc.) requirements to meet MOEs.

3   Procedure for Decisions Trade-Off

All information and inputs of our MOO problem we derive from the Model-Based Architectural Framework incorporated with PESTEL and FURPSE analyses, to which we refer in all the iterative analyses phases (for operational, functional, structural views).

3.1   Operational Analysis Phase: Derivation of Invariants

The system shall be correctly integrated within its environment to ensure that its mission is successfully accomplished. As explained in [1], we start off by defining and categorizing numerous stakeholders (direct and indirect) and their inherent complexities within the PESTEL frame (Political, Economic, Social, Technological, Environmental and Legal) [INCOSE], including Regulatory/Standardization, Competition and Organizational axes, as indispensable. It is further interesting to consider any aspects from System of Systems (SoS) perspective (Tables 2 and 3).

Table 2   Macro needs refinement

N2.1	Operators want an automatic adjustment of resources allocation for traffic growth and de-growth
N2.2	Operators want maximal availability and speed for huge traffic rates
N2.3	Operators do not want to feel the physical limitations of the system
N2.4	Operators want predictable behavior of network functions
…	

Table 3   Example: parameters

N	General traffic model parameters	Measure units	Values
TM1	Min. number of CPU cores	Unit	
TM2	Min. number of core network elements	Unit	
TM3	Min. dynamic memory	GB	
TM4	Min. direct attached storage	GB	
TM5	Min. block storage (Cinder)	GB	
TM6	Min. object storage	GB	
TM7	Min. estimated performance	Msgs/Sec.	
TM8	Max. response/ressources allocation latency	Sec.	<2
TM9	Max. failure latency	Sec.	
TM10	Max. failures/incidents per year	Percentage	
	Min. availability	Percentage	
	….		

Reflecting the broad range of stakeholders needs, decisions trade off analyses (according to market and technological criteria, as of primary importance compared to other dimensions and their development dynamics) need to be performed iteratively throughout the system lifecycle phases. We characterize them in terms of desired software (functional) architecture properties, as FURPSE (Functionality, Usability, Reliability, Performance, System Maintainability, and Evolution) [ISO/IEC 9126 norm].

Operational analysis defines, in a non-ambiguous way, the system's Stakeholders, Needs, Mission, external Interfaces, Contexts, Uses Cases and Scenarios. Operational invariants are inputs for functional analysis and related decisions tradeoffs.

Based on Use Cases and Scenarios analysis, the macro functions are first defined, which becomes the basis for Functional Breakdown Structure (FBS), that represents the static functional view. Macro functions, hence the FBS, are consequently refined into micro functions and further on completed by behavioral and functional modes.

3.2 Functional Analysis Phase

Every function has one or several modes (normal = nominal, operational or degraded) of functioning. The static functional view analysis shall be continued with the dynamic view analysis, defining all possible functional behaviors and ways. In the figure below we present our functional and structural model example. On the left side are grouped the Macro functions organized in FBS and on the right side their corresponding components organized in PBS (Product Breakdown Structure). <Allocated to> arrows show the Functions to Components allocation process (see Fig. 2).

Evidently, there are numerous diverse external interfaces depending on the IMS life cycle phase. For clarity, in this paper, we would focus only on those that are prior within the given context.

4 Optimization Problem Inputs and Objectives

The principle inputs for the MOO problem are based on needs of the direct stakeholders, listed below from [1]. Information resulting from updated market studies shall be permanently taken into consideration during this step.

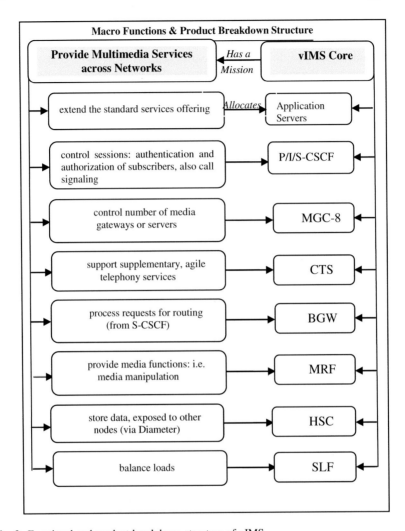

Fig. 2 Functional and product breakdown structure of vIMS

In this paper, any example is only illustrative and not exhaustive or Alcatel-Lucent's data representative.

4.1 Elicitation of Needs: Inputs of MOO Problem

As a result of refinement procedure, each Macro need is present by a set of relevant Micro needs. One example is presented below:

4.2 Analysis of Inputs

Parameters, variables, constraints that are the essential inputs for the MOO model, shall be derived from the needs analysis. The refinement shall be continued until reaching the level of clear, precise, measurable, and quantifiable needs, which are further translated into requirements. Hence, we derive the specific sets of parameters that characterize: traffic profiles, service user parameters, parameters specific to Network Elements, etc. For an advanced analysis, the SLA parameters (i.e. descriptive of reliability, availability, security, etc.) [Ref. ETSI NFV high level requirements] shall also be taken into consideration. The list below is partial, for an illustrative only.

4.3 Objectives of Optimization Problem

At first hand, some requirements become optimization objectives. For instance, the requirement *IMS shall not be expensive* yields an objective that is *Minimize CapEx*. Another requirement *IMS shall not be expensive at use* yields another objective: *Minimize OpEx*.

From a high level trade-off perspective, as the picture below illustrates (see Fig. 3), the NFV strategic objectives are to be achieved through the chain of NFV enablers: dynamic resiliency (scaling (up/down) and (de-) growth), end-to-end solution efficiency (capacity and performance), orchestration (resources allocation/ placement), automation (self-regulation), etc.

Maximization of Automation and Minimization of Customization shall lead to Minimization of Services Efforts and Time. From the other hand, Maximization of Automation and Minimization of Customization Maximizes the Resources Allocation/Usage Orchestration, which in its turn contributes to Maximize the Scaling and Growth possibilities. This chain of objectives Maximizes the Absolute Performance (to attain maximal capacity) and Relative Performance (to attain maximal efficiency), which in sum contributes to the Minimization of Resources Waste. Both Resources and Services efforts and time wastes minimization contribute to Maximization of System Autonomy and Minimization of Costs of Ownership. From the other hand, Maximization of Scaling and Growth leads to Maximization of Capacity.

The goal of the NFV enablers is to make the best, most optimal usage of the system and cloud resources. Network functions have different resource requirements, some with considerably larger capacity requirements, which are the number of comprising virtual machines (VMs), each with a specific CPU, memory estimations. Either by adding or removing resources (additional capacities) or adding or deleting instances, Operators need to optimize the network usage according to different traffic profiles, user data.

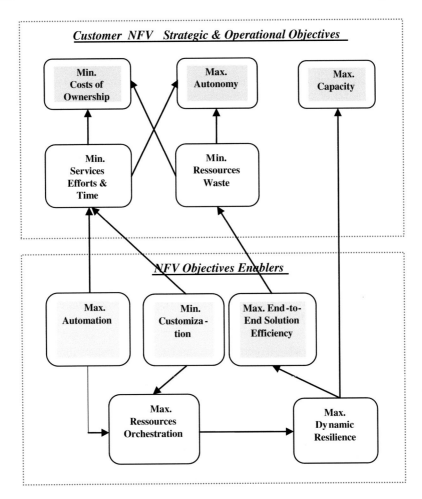

Fig. 3 Direct market *strategic and NFV objectives*

Moreover, proactive and predictive analytics is to be embedded in this procedure, comprising knowledge upon services, traffic scenarios. It shall help to monitor and manipulate the present and future states of resources. Decisions shall be made instantly from an uncertainty perspective, while ensuring the minimal efforts, costs, maximal availability, etc. Apart resources estimations, their optimal placement on a dedicated infrastructure is another key objective of orchestration.

Among many functional and performance constraints to be taken into consideration, there are also NFV challenges, as amplified even more with virtualization and cloudification: Security, Reliability, Serviceability, etc.

Important is to note, that optimization becomes more difficult in a geographically distributed datacenters, when additional constraints need to be satisfied, as performance related, regulatory, etc.

Further on, each Maximization and Minimization objective shall be represented through mathematical equations and simplified. At the end, this procedure shall serve for simulations.

Perspectives Discussion

In this paper, at the basis of our previous study results [1], we proposed a baseline for decisions trade-off and functional analysis policy. We explain the importance and usefulness of our approach for anticipating and managing any evolutions from strategic and system engineering perspectives. We constituted our approach based on metaphoric probes and proposed: (1) Model-Based Architectural Framework incorporated with PESTEL and FURPSE analyses frames to model the system's views (operational, functional, physical), environment, direct market specifics in a structured, interrelated and systemic manner; (2) Multi-Objective Optimization technique for definition of functional and decisions options and further simulations.

For this, we further aim to (a) define the optimal scaling and growth points according to different traffic profiles, (b) derive the factors/rules for each IMS component, (c) figure out resource requirements from the scaling rules, (d) apply the theoretical placement and sizing rules to produce a set of optimal configurations for virtual IMS in the cloud environment, and at the end, to (e) confront these optimum configurations with real case tests.

For the future research, we propose to enlarge our NFV perspectives and continue the study for dynamically *self-optimizing virtual networks*. For instance, in case of a maximal automation, many questions, as: virtual machines placement, behavior driven instantiation and scaling, hardware failure cases, etc. shall be evaluated carefully. Moreover, it shall be vitally useful to create a set of rules to govern the IMS universe.

Lastly, we may like to reflect upon the infrastructural resource usage consensus, when deploying multiple virtual IMS systems within a shared cloud. In this perspective, last phase of optimization may be envisaged in the sense of game-theoretical equilibrium.

References

1. Gevorgyan, A., Spencer, P.: Operational Analysis of Virtual IP Multimedia Subsystem Through a Model-Based Architectural Framework accepted for publication in CSDM 2015
2. Bartolomei, J.E., Hastings, D.E., de Neufville, R., Rhodes, D.H.: Engineering Systems Multiple-Domain Matrix: an organizing framework for modeling large-scale complex systems, MIT, Accepted 24 February 2011, Published online 10 October 2011 in Wiley Online Library (wileyonlinelibrary.com)
3. Krob, D.: Eléments d'architecture des systèmes complexes. In: Gestion de la complexité et de l'information dans les grands systèmes critiques, pp. 179–207. CNRS Editions (2009)
4. Krob, D.: Enterprise Architecture, Modules 1–10, Ecole Polytechnique, 2009–2010 (personal communication)
5. Estefan, J.A.: Survey of Model-Based Systems Engineering (MBSE) Methodologies, Report of INCOSE MBSE Focus Group, Rev. A, 25 May 2007

6. ETSI GS NFV: Network Functions Virtualization (NFV), Use Cases Specification 004 V1.1.1"
7. INCOSE: Systems Engineering Vision 2020 (2007)
8. Chalé Góngora, H.G., Gaudré, T., Tucci-Piergiovanni, S.: Towards an Architectural Design Framework for Automotive Systems Development, CSD&M 2013
9. Chalé Góngora, H.G., Dauron, A., Gaudré, T.: A Commonsense-Driven Architecture Framework. A Car Manufacturer's (naïve) Take on MBSE. INCOSE (2012)
10. Berrebi, J., Krob, D.: How to use systems architecture to specify the operational perimeter of an innovative product line. INCOSE 22(1), 84–99 (2012)
11. Doufene, A., Krob, D., Chalé Góngora, H.G., Dauron, A.: Model-Based operational analysis for complex systems—a case study for electric vehicles. INCOSE 24(1) (2014)
12. Zachman, J.A.: A framework for information systems architecture. IBM Syst. J. 26(3), 276–292 (1987)
13. Friedenthal, S., Moore, A., Steiner, R.: A practical guide to SysML—The Systems Modeling Language, Morgan Kaufmann (2008)
14. Honour, E.C.: Understanding the value of systems engineering. INCOSE (2004)
15. INCOSE: Systems Engineering Handbook. A guide for system lifecycle processes and activities. International Council on Systems Engineering (INCOSE), San Diego, CA (2010)
16. PESTEL: PESTEL analysis of the macro-environment. Oxford University Press (2007)
17. Penalva, J.M.: La modélisation par les systèmes en situations complexes, PhD Thesis, Université de Paris 11, Orsay, France (1997)
18. http://www.3gpp.org/
19. http://dodcio.defense.gov/Portals/0/Documents/DODAF/DoDAF_v2-02_web.pdf
20. https://www.gov.uk/mod-architecture-framework
21. Björnander, S., Grunske, L.: Architecture Description Languages for Automotive Systems—A Literature Review. Technical Report: C4-01 TR M49, 30 July 2008
22. Weilkiens, T.: Systems Engineering with SysML/UML—Modeling, Analysis, Design. Morgan Kaufmann Publishers (2008)
23. Saaty, T.L., Vargas, L.G.: Models, Methods, Concepts and Applications of the Analytic Hierarchy Process, International Series in Operations Submitted for publication in the Journal of Systems Engineering (April 2011) Research and Management Science. Springer (2000)
24. Cisco: Evolution of the Mobile Network White Paper (2010) http://www.cisco.com/c/en/us/solutions/collateral/service-provider/mobile-internet/white_paper_c11-624446.html
25. 4G Americas: Bringing Network Function Virtualization to LTE, White Paper (2014)
26. Kang, K.C., Cohen, S.G., Hess, J.A., Novak, W.E., Peterson, A.S.: Feature-Oriented Domain Analysis (FODA) Feasibility Study. Technical Report, DTIC Document (1990)
27. Pohl, K., Bockle, G., Van der Linden, F.J.: Software Product Line Engineering: Foundations, Principles, Techniques. Springer (2005)
28. Roos-Frantz, F., Benavides, D., Ruiz Cirts, A., Heuer, A., Lauenroth, K.: Quality—Aware Analysis in Product Line Engineering with Orthogonal Variability Model. Softw. Qual. J. (2011)
29. Becker, M.: Towards general model of variability in product families. Workshop on Variability Management, Groningen, The Netherlands, pp. 19–27 (2003)
30. Possomps, T., Dony, C., Huchard, M., Rey, H.: Tibermacine, C., Vasques, X.: A UML profile for feature diagrams: initiating a model driven engineering approach for software product lines. Journe Lignes de Produits. pp. 59–70 (2010)
31. Von der Masen, T., Lichter, H.: RequiLine. A Requirements Engineering Tool for Software Product Lines. Springer (2004)
32. Halmans, G., Pohl, K.: Communicating the variability of a software-product family to customers. Softw. Syst. Model. 2(1), 15–36 (2003)
33. Tessier, P., Servat, D., Gerard, S.: Variability management on behavioral models. VaMoS Workshop, pp. 121–130 (2008)

34. Gmez, A., Ramos, I.: Automatic Tool Support for Cardinality-Based Feature Modeling With Model Constraints for Information Systems Development. Information Systems Development, pp. 271–284. Springer (2011)
35. Salinesi, C., Mazo, R., Diaz, D., Djebbi, O., Lora-Michels, A.: Constraints: The Core of Product Line Engineering. Research Challenges in Information Science (RCIS), pp. 1–10 (2011)
36. Streitferdt, D., Riebisch, M., Philippow, K.: Details of formalized relations in Feature Models Using OCL. In: Proceedings of the 10th International Conference and Workshop on Engineering of Computer-Based Systems, pp. 297–304 (2003)
37. Burns, E.V., Duggal, D., Kraft, F.M., Matthias, J.T., McCauley, D., Palmer, N., Pucher, M.J., Silver, B., Swenson, K.D.: Taming the Unpredictable. Real World Adaptive Case Management: Case studies and Practical Guidance (2011)
38. Near Optimal Placement of Virtual Network Functions, INFOCOM 2015
39. Online allocation of virtual machines in a distributed cloud, INFOCOM 2014
40. Joint static and dynamic traffic scheduling in data center networks, INFOCOM 2014
41. Network aware resource allocation in distributed clouds, INFOCOM 2012
42. Optimizing data access latencies in cloud systems by intelligent virtual machine placement, INFOCOM 2013

Portfolio Decision Technology for Designing Optimal Syndemic Management Strategies

Matteo Convertino and Yang Liu

Abstract Cholera is an infectious disease responsible for roughly 3–5 million morbidities and 100,000–120,000 mortalities every year at the global scale. Frequent cholera outbreaks in the recent history suggest unresolved inefficiency issues with regards to cholera prevention and intervention strategies. Guidelines of the World Health Organizations (WHO) advise country governments facing threats of cholera epidemics to prevent and control potential outbreaks by developing effective sanitation, proper waste management strategies and vaccination campaigns. These controls do not envision any focus on environmental determinants of cholera outbreaks. Failing to select the most appropriate prevention and intervention strategies at the health management scale based on public health, environmental, and social determinants is the fundamental cause for the low effectiveness of cholera outbreak containment strategies. This study targets this inefficiency via the creation of a model-based technology that detects the optimal combination of outbreak controls which minimize the number of cases at the system scale. As a case study we consider cholera but the model can be applied to any syndemic and/or complex diseases affected by natural and human systems. The technology is based on the integration of an epidemiology model that processes public health information and predicts population dynamics during the epidemic, an environmental model that predicts environmental fluxes (i.e., hydrologic fluxes) and a mobility model that predicts human fluxes. Results from the physical based model feeds a Portfolio Decision Model (PDM) that is composed by a Multi-Criteria

M. Convertino (✉)
HumNat Lab, Division of Environmental Health Sciences and PH Informatics Program, School of Public Health, Institute on the Environment and Institute for Engineering in Medicine, University of Minnesota, Twin Cities, MN, USA
e-mail: matteoc@umn.edu

Y. Liu
HumNat Lab, Division of Environmental Health Sciences and PH Informatics Program, School of Public Health, Twin Cities, USA
e-mail: liux3204@umn.edu

© Springer International Publishing Switzerland 2016
M.-A. Cardin et al. (eds.), *Complex Systems Design & Management Asia*,
Advances in Intelligent Systems and Computing 426,
DOI 10.1007/978-3-319-29643-2_17

223

Disease Analysis (MCDA) and a Pareto optimization model. The MCDA model is used for the static evaluation of the feasible controls at the smallest community scale; the Pareto optimization detects the most appropriate control strategy rather than one single control alternative. Preliminary applications of the model applied to the great Kolkata ecosystem shows an average 35 % decrease in incidence for the portfolio versus the monocontrol scenario. Acknowledging spatial sensitivities in the epidemiological dynamics, PDM benefits public health management concerned with multiple populations with heterogeneous dependencies occurring simultaneously. PDM considers public health management scales and optimizes the distribution of economic resources for minimizing the risk of infection at the system scale. A major innovation is constituted by the explicit consideration of environmental dynamics, global sensitivity and uncertainty analysis, and MCDA that is particularly relevant for bringing together biophysical factors and stakeholder preferences in the decision making process. The model can be extended from one disease to syndemics linked together by common socio-environmental drivers or the structure of the natural-human systems responsible for their spreading.

1 Introduction

1.1 Rationale

The key of understanding system dynamics and formulating reliable predictions lies in the elucidation of effective transmission networks underlying observed health patterns. For instance, these patterns can be outbreak patterns of cholera and affine waterborne syndemics, where syndemics are defined in terms of co-causality of socio-environmental factors for two or more diseases where causality can be lagged in time for different diseases. However, the knowledge of effective transmission networks is not enough for disease analysis because proper management place and their global effectiveness should be known. Considering cholera, Vibrio cholera spreads from community to community through pathways represented by transmission networks. This network reflects both environmental transmission through a hydrological network and person-to-person transmission through human mobility networks between communities. For any given system, the overall structure of transmission networks influences how frequently and quickly a disease spreads from one community to another, as well as the efficacy of different control strategies (both for prevention and intervention).

The key issue of disease management is to find out the best control strategies that minimize the overall disease incidence over space and time. For syndemics where multiple causal factors and outcomes exist, the systemic risk [10] can be introduced as the outcome variable to consider and to minimize (Fig. 1). In this view the

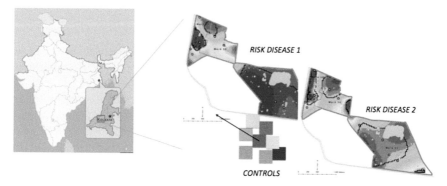

Fig. 1 Schematic of the syndemic management problem (multi-disease/-control), case study and desired output of the proposed portfolio management technology. The *top left* plot map is representing the area considered as a case study (large Kolkata macroregion); the *top right* plot is considering multiple diseases co-occurring in the same area and related to each other by common environmental causes and/or controls. The *bottom* plot is the auspicated output of the portfolio technology that provides optimal management strategies (multiple controls for multiple diseases) for a given available budget. Points on the portfolio frontier (*light green*) maximize the systemic value (complimentary of the systemic risk) of the population considered. The systemic value is a multi-criteria value in which disease incidence is minimized and other factors—social, environmental and economical factors—may be minimized or maximized. The systemic value is determined using a MCDA model and the portfolio solution is determined via a Pareto optimization algorithm

systemic risk represent the likelihood of observing multiple diseases simultaneously. Particularly for syndemics, this management problem is a non-trivial portfolio problem where potential Pareto optimal solutions should be found via a stochastic exploration of all possible management strategies and outcome [7, 13]. Often, communities respond locally to an outbreak without considering the interconnections between communities (e.g. by containment of infected individuals). For instance, controls that act only in disease hotspots are not likely optimal controls because they are point-based actions that do not consider the whole paths of factors causing cholera outbreaks. However, a control that works well in one location may not be as effective in another because of its position in the network, and more importantly at the system scale in terms of infection risk abatement. Previous work has shown that controls that leverage information about the transmission network structure at the system scale to prevent or respond to outbreaks outperform those that do not.

Additionally, some controls may be relevant at one or more scales depending on the importance of factors they control at the scales considered. Physical networks of ecosystems and related control strategies affect the causal transmission network of interacting factors determining cholera cases. The transmission network can in fact be highly heterogeneous and the determinants of infections in different areas vary.

1.2 Proposed Approach

In order to incorporate values and objective probability simultaneously in a strategic decision making approach for disease management we propose to build a portfolio decision making technology based on a portfolio decision model (PDM) (Fig. 1). PDM clearly underlines the importance of coupling choice and technicality of an effective decision making process. Data and/or model-based disease predictions will serve as the input to the portfolio decision model (PDM) that consists of a multi-criteria decision analysis (MCDA) tool and a Pareto optimization model (Fig. 2).

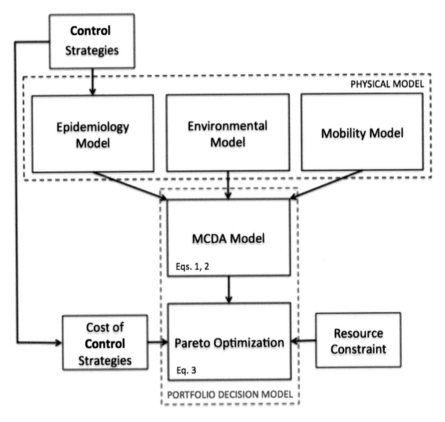

Fig. 2 Framework of the portfolio decision model technology. The portfolio model is combining a MCDA model and a Pareto optimization model to identify the best set of disease controls (prevention and intervention controls) that maximize population values at the system scale (health is an objective) for a given budget. The MCDA scores each control considering social, environmental and economical criteria and the Pareto optimization finds the best set of controls. The novelty is the simultaneous integration of a biophysical model (e.g. for cholera see You et al. [15] and Convertino et al. [9]) considering the environmental and epidemiological dynamics and a decision science model that can include stakeholder preferences for optimizing the decision making process related to syndemic management

MCDA is chosen instead of other quantitative decision science methods—such as multi-objective evolutionary algorithms [11]—because our vision to pursue the PDM of Convertino and Valverde [7], and because policymakers are more concerned with considering stakeholder values and technical information comprehensively into one unique objective function to optimize and use for system design. Multi-objective evolutionary algorithms are more appropriate when there are few systems' drivers under consideration and multiple outcomes of interests; typically, these methods consider multiple objectives as independent variables and highlight their trade-off after their simultaneous minimization and maximization leaving decision makers with high no indication of an optimal solution. Yet, the PDM of Convertino and Valverde [7], here used in the modeling, is focused on the detection of the optimal system design—considering spatio-temporal dependencies of criteria and budget constraint—versus highlighting optimal tradeoffs between design objectives.

2 Methods

2.1 Multi Criteria Decision Analysis Model

The MCDA model used for evaluating each management plan for human and natural assets is a linear Multi Attribute Value Theory (MAVT) model [12]. A management plan, or strategy, is a set of prevention and intervention controls of multiple diseases in multiple communities; we refer to populations as assets for which health is part of a population value to maximize. The MCDA model ranks management controls by scoring them as linear combination of criteria values and criteria weights. Criteria are social, environmental and economical criteria; social criteria consider population health outcomes such as disease incidence and geographic range. The MCDA model calculates the local value for management plan $\underline{R}$ with respect to asset j in management area m at each of the simulation. We assume implicitly the dependence on time of the predicted values. The MCDA value that is ultimately the value of assets given a management plan is:

$$V_{m,j}(\underline{R}) = \sum_k w_{j,k} x_{j,k,m}(\underline{R}). \tag{1}$$

The weights $w_{j,k}$ are assigned by stakeholders, and correspond to the relative importance of each criterion for the local value of management controls. Here, we utilize an illustrative set of weights elicited according to our expert knowledge about the problem. The MCDA model takes as input spatially explicit predictions of local- and community-scale criteria from biophysical models that include environmental, social and epidemiological dynamics. Each management control is evaluated every year for each feasible management plan, for the planning horizon considered. We may assume the projected climate proceeding according to the A1B scenario. Criteria values $x_{j,k,m}$ depend on the whole management plan $\underline{R}$. These criteria are maximized or minimized for rescaling criteria values to increasing or decreasing value functions in a range dictated by the maximum and the minimum criteria values. For example, pathogen and/or disease suitability is a criterion that depends on local heterogeneities, but also on the whole configuration of the ecosystem. The MCDA value of assets does not take into account the likelihood of success of the management plan. The likely success of each management control is evaluated by both considering the vulnerability of each asset at the global scale and the local effectiveness of each management control. These factors multiply the MCDA value to obtain the expected value of assets.

2.2 Expected Local Value of Human Population

The expected local value in area m for asset j and for management plan $\underline{R}$ at each of the analysis is given by:

$$V^*_{m,j}(\underline{R}) = (1 - v_j(\underline{R}))f_{i(j)}R_{i(j),m}V_{m,j}(\underline{R}). \tag{2}$$

The value of a management plan is adjusted by the probability of success given by the vulnerability of each assets at the global scale ($v_j(\underline{R})$) and the effectiveness of a management plan ($f_j = f_{i(j)} R_{i(j), m}$), considering the management interventions selected locally in the management areas ($R_{i(j),m}$). The expected values are calculated for both the MCDA-based method and the PDM-based selection methods.

2.3 Expected Global Value of Human Population

The global value of human-natural assets, $V_T(\underline{R})$, of a management plan (i.e., a set of management alternatives at the community scale) is calculated as a Euclidian distance.

We selected the Euclidian distance because we hypothesize the ecosystem services—where population health is included—as additive services. Ecosystem services are the benefits that natural and human assets can get by controls that preserve the habitat they use. The Euclidian distance allows one to analyze easily the contribution of both factors. A simple sum of natural and human values produces the same Pareto frontiers bur rescaled vertically in values. The Euclidian distance is used because the space of ecosystem services is often seen as a multi-dimensional space, and thus the distance is the most proper function to consider services together.

The values of human and natural assets weighted by stakeholder preferences at the global scale, $V_H(\underline{R})$ and $V_N(\underline{R})$, are the two components of the distance. These values are $V_H(\underline{R}) = \sum_m \sum_j V_{H\,m,j}^*(\underline{R})\, w_j$ and $V_N(\underline{R}) = \sum_m \sum_j V_{N\,m,j}^*(\underline{R})\, w_j$ respectively. Because of the absence of a careful stakeholder preference elicitation, in this study we assume the preferences w_j to be homogeneously distributed among assets because the lack of a stakeholder preference elicitation effort. The elicitation of preferences should be performed in real-world applications of the model. The global value of a management plan calculated for both the MCDA-based and PDM-based selection method at each year of the analysis is given by:

$$V_T(\underline{R}) = \sqrt{\sum_{m=1}^{M} \sum_{j=1}^{J} (V_{m,j}^*(\underline{R})w_j)^2} = \sqrt{V_N(\underline{R})^2 + V_H(\underline{R})^2}. \tag{3}$$

2.4 Pareto Optimization Model

The Pareto optimization model assumes that the global value of assets for each management plan is a multi-objective function that maximizes (Eq. 3). We assume mutually independent management controls under uncertainty. The optimization is a linear mixed-integer optimization algorithm that explores all possible combinations of management controls (for each asset, one at a time), with their expected local value and cost at the community scale. The maximization of the global value is performed with and without the constraint of the available budget B of the community, for each simulated time period. In the constrained case, the cost of the management plan, $C(\underline{R}) = \sum_{m=1,M} \sum_{j=1,J} C_{m,j}(R_{i(j),m})$, cannot exceed the budget B. In the case of the Pareto optimization unconstrained by budget, if $V_T(\underline{R}_1) \geq V_T(\underline{R}_2)$ and $C(\underline{R}_1) < C(\underline{R}_2)$, then the portfolio solution $\underline{R}_1$ dominates $\underline{R}_2$. Thus, all the management controls in $\underline{R}_1$ are selected. In the budget constrained Pareto optimization, if $V_T(\underline{R}_1) \geq V_T(\underline{R}_2)$ then the portfolio combination $\underline{R}_1$ dominates $\underline{R}_2$.

Optimal management plans are defined as the Pareto-efficient solutions along the Pareto frontiers calculated by the optimization model. Pareto frontiers can also show the relative importance of natural and human assets. In our budget constrained case, we assume that natural and human assets are equally important at the community scale. Thus, the ratio $V_H(\underline{R})/V_N(\underline{R})$ is equal to one. Unaffordable plans are those for which the cost is higher than the available budget, and with global value of assets lower or higher than the Pareto optimal plans. These portfolio solutions are above the Pareto frontiers unconstrained to the budget available. Affordable plans are those whose total cost is lower than the budget, but they are suboptimal in term of global value ($C(\underline{R}_1) < C(\underline{R}_2)$ and $V_T(\underline{R}_1) < V_T(\underline{R}_2)$). The number of management interventions contained in any Pareto optimal solution is equal to the product of the number of assets considered and the number of management areas.

2.5 Global Sensitivity and Uncertainty Analyses

To consider uncertainty in the selection of management plans we add a truncated Gaussian noise term μ in the local values of assets determined by the MCDA model (Eq. 1). The noise is assumed to fall in the range [−0.05, 0.05], with mean and standard deviation equal to zero and ½, respectively. The same white noise has been assigned to the cost of the management controls considered. Hence we both considered uncertainty in the benefits and costs of the management controls. The truncated noise is included to add further uncertainty which may be related to other factors not considered in the biophysical models and portfolio models: for example pathogenic species changes in habitat preferences that change their habitat suitability, and changes among pathogens which may change the asset correlation based on such inter-relationships. The truncated Gaussian noise is also taking care of all deterministic factors included in the portfolio such as the effectiveness factor, stakeholder preferences, and the extent of communities.

Thirty Monte Carlo simulations are generated by considering such uncertainties in local values and costs of management controls. Thirty simulations are enough to capture the variability of the output based on our previous studies of the uncertainty in the biophysical models [1–5, 7, 8] by performing a global sensitivity and uncertainty analysis [14].

3 Results and Discussion

Considering the environmental and social criteria leading to cholera epidemiological dynamics the MCDA generates the potential systemic risk for each subpopulation as a function of local controls. Each systemic risks and their corresponding budget constraint can be plotted in a two dimensional space, where each points represents a collection of five alternatives chosen for each population. The Pareto optimization model selects the point minimizing systemic risks at each budget constraint level. By connecting these points, we discovered the Pareto frontiers, which are the best possible ways to allocate resources under each scenario. In Fig. 3, we plotted the Pareto frontiers of seven monocontrol strategies and the portfolio management method. The shapes of these Pareto frontiers are close to exponential decay.

When no action is taken, the systemic risk of cholera epidemics is 3.28. Intuitively, when budget constraint goes up, systemic risk goes down. Overall, the green line is the lowest at budget constraint levels between 0 and 88. Further increasing the budget constraint levels through the portfolio management method does not change the optimal collection of alternatives selected for each populations under consideration. At budget constraint of 88, the overall systemic risk is 0.32.

Among monocontrol strategies, water filtration can achieve the closest systemic risk level to the portfolio management method, at 0.33. However, the corresponding

Pareto Optimal Frontiers

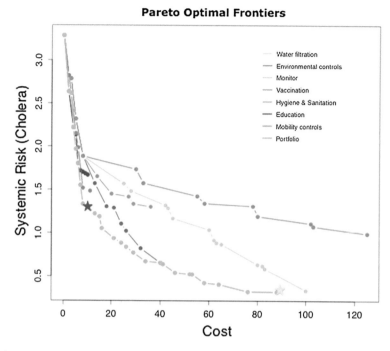

Fig. 3 Pareto Frontiers of monocontrol and portfolio management strategies. The primary strategy under consideration is depicted by colors. *Dots* represent a collection of alternatives that form a strategy outcome. When there's no strategy outcome corresponding to a certain budget constraint level, the Pareto optimization model forces the system to select the first strategy outcome with a lower budget constraint. The *red* and *yellow stars* are for the lowest and highest resource portfolio solution that minimize and maximize the diversification (of risk and management strategies), respectively (Fig. 4)

cost is 100. This monocontrol strategy is strongly dominated by the portfolio management method. The most inefficient monocontrol strategies is vaccination. For most budget constraint levels, its systemic risk results are the highest. At the highest budget constraint level, which equals 125, the systemic risk is still as high as 0.99. Overall, Fig. 3 shows the capacity of the portfolio management method to significantly improve the quality of decision-making processes with regards to cholera prevention and control. It is very interesting to note that both systemic risk and cost of disease management are lower for the portfolio solution than mono-control solutions (Fig. 4). In analogy to the financial market, this is associated to the ability of the portfolio to diversify the risk optimally (because of the Pareto opti-mization model) across all feasible disease management alternatives.

The results cannot be validated due to the absence of long records of disease management strategies. In this regard, systemic surveillance is not yet in place in the majority of countries. However, it is certainly true that measures of effectiveness

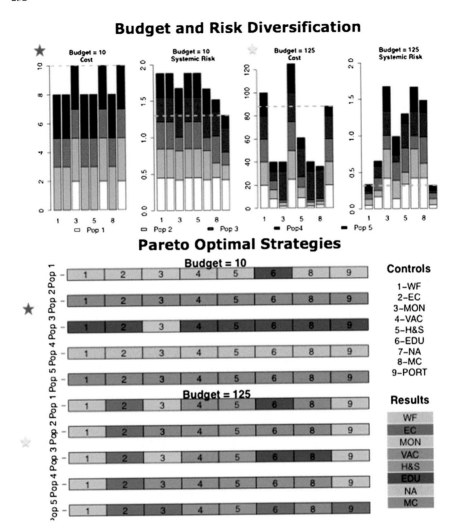

Fig. 4 Portfolio management benefits and diversification chart. In the *top* plot the minimum and maximum budget constraint levels are considered: 10, and 125. For each budget constraint level, spending and risks of cholera epidemics are attributed to each of the five populations under consideration. The *green dashed line* represents the systemic risk and cost achievable via the portfolio management method. Both risk and cost are lower for the portfolio solution than monocontrol solutions. In the *bottom* plot (diversification chart) the *colors* represent the alternatives selected for each of the five populations under consideration and the corresponding budget constraint

of disease controls are available, yet a comparative analysis can be undertake using in silico simulations. Global sensitivity and uncertainty analysis guiding the Monte Carlo simulations allows one to explore the stability of prediction for any mono-control and portfolio solution.

The study shows the impact of a systemic approach to health management using the portfolio management model as opposed to more traditional control strategies that are focused on one single disease only (e.g. vaccination). In particular, results show that environmental control (here in the form of hydrological management) of the natural and built environment is the most effective control on populations affected by cholera. It should be noted though that the hydrological contro of the environment, for instance to control floods via civil and environmental engineering infrastructure, has a huge effects on multiple diseases (waterborne and water-based in particular) and population outcomes simultaneously.

Lastly, we believe that the portfolio model is also important for showing gaps in outcomes from the status quo. This evidence from gaps can trigger stakeholder shifts in decision making oriented toward more systemic management solutions that optimize system structure for achieving the desired ecosystem services.

4 Conclusions

PDM allows stakeholders to use information from biophysical models and field data integrated to subjective assessment of criteria distributions and preferences. The model is used to formulate optimal and spatially explicit decision strategies for disease management in the management horizon. Moreover, PDM is built as a dynamic tool to use in an adaptive management context for almost real-time decision making in presence of surprises, new information, budget revisions, and/or preference shifts.

Results show that population health is significantly improved by the portfolio management method compared to monocontrol strategies. Spatial sensitivities based on health management scale are both beneficial and necessary. The traditional "one-for-all" approach that assumes perfect homogeneity of the ecosystem is inefficient as it diffuses efforts without sound reasoning. It treats target population as identical, thus often provides insufficient assistance for high-risk areas and excessive assistance for low-risk areas. The new approach in this study dramatically improves the efficiency of intervention activities by taking local conditions into consideration. It tailors the local intervention activities based on its potential of cholera epidemics with the overall goal of minimizing systemic risk. Our PDM shows a minimum risk reduction of 35 % through the portfolio management method.

The theoretical framework of this study is extremely pragmatic. It uses all available information, even when it is incomplete since such information is essential to the success of public health management. It may as much as double the complimentary of morbidity and mortality relative to using (admittedly incorrect) assumptions commonly invoked. Furthermore, the incompleteness is compensated by the iteration of the PDM itself due to its dynamic nature. The selection of the optimal set of prevention and intervention controls can be carried out using the proposed probabilistic portfolio decision model, which is updated as soon as new information on infectious disease epidemiology or stakeholders preferences become available. The general setup of the PDM in the context of this study makes it a particularly powerful tool while solving challenges posed by environmentally sensitive infectious diseases. Furthermore, the inclusion of GSUA and MCDA makes the overall approach highly relevant for bringing together biophysical factors and stakeholder preferences into the decision making process. In the future, we expect to see similar methods to be adopted for different environmental infectious diseases at different locations to further test the robustness of the PDM.

References

1. Chu-Agor, M.L., Muñoz-Carpena, R., Kiker, G.A., Emanuelsson, A., Linkov, I.: Exploring sea level rise vulnerability of coastal habitats through global sensitivity and uncertainty analysis. Environ. Model Softw. **26**, 593–604 (2011)
2. Chu-Agor, M.L., Muñoz-Carpena, R., Kiker, G.A., Aiello-Lammens, M., Akçakaya, R., et al.: Simulating the fate of Florida Snowy Plovers with sea level rise: exploring potential population management outcomes with a global uncertainty and sensitivity analysis perspective. Ecol. Model. (2012)
3. Convertino, M., Kiker, G.A., Muñoz-Carpena, R., Fischer, R.A., Linkov, I.: Scale- and resolution-invariance of suitable geographic range for shorebird metapopulations. Ecol. Complex. (2011). doi:10.1016/j.ecocom.2011.07.007
4. Convertino, M., Kiker, G.A., Chu-Agor, M.L., Munoz-Carpena, R., Martinez, C.J., et al.: Integrated Modeling to Mitigate Climate Change Risk due to Sea Level Rise of Imperiled Shorebirds on Florida Coastal Military Communitys, NATO Book. In: Linkov, I., Bridges, T. (eds.) Climate Change: Global Change and Local Adaptation (2011)
5. Convertino M., Muñoz-Carpena R., Kiker G.A., Chu-Agor M.L., Fischer R.A., et al.: Epistemic uncertainty in predicted species distributions: models and space-time gaps of biogeographical data. Ecol. Model. (2011)
6. Convertino, M., Chu-Agor, M.L., Fischer, R.A., Kiker, G.A., Munoz-Carpena, R., et al.: Coastline fractality as fingerprint of scale-free shorebird patch-size fluctuations due to climate change. Ecol. Process. (2012)
7. Convertino, M., Valverde Jr, L.J.: Portfolio decision analysis framework for value-focused ecosystem management. PLoS One **8**(6), e65056 (2013). doi:10.1371/journal.pone.0065056
8. Convertino, M., Chu-Agor, M.L., Baker, K., Linkov, I., Munoz-Carpena, R.: Untangling drivers of species distribution models: global sensitivity and uncertainty analysis of MaxEnt. Environ. Model. Softw. (2014)
9. Convertino, M., Liang, S., Arabi, M., Morris, S.: Unveiling the spatio-temporal cholera outbreak in cameroon: a model for public health engineering. BMC Infect. Dis. (2015) (in press)
10. Haldane, A.G., May, R.M.: Systemic risk in banking ecosystems. Nature **469**, 351–355 (2011). doi:10.1038/nature09659
11. Kasprzyk, J.R., Reed, P.M., Characklis, G.W., Kirsch, B.R.: Many-objective de Novo water supply portfolio planning under deep uncertainty. Environ. Model Softw. **34**, 87e104 (2012)
12. Linkov, I., Moberg, E.: Multi Criteria Decision Analysis: Environmental Applications and Case Studies. CRC Press (2012)
13. Salo, A., Keisler, J., Morton, A.: Portfolio Decision Analysis, Improved Methods for Resource Allocation, 1st edn., XV, p. 409 (2011)
14. Saltelli, A., Ratto, M., Andres, T., Campolongo, F., Cariboni, J., Gatelli, D., et al.: Global Sensitivity Analysis: The Primer. Wiley (2004)
15. You, Y.A., Ali, M., Kanungo, S., Sah, B., Manna, B., et al.: risk map of cholera infection for vaccine deployment: the eastern Kolkata case. PLoS One **8**(8), e71173 (2013). doi:10.1371/journal.pone.0071173

Using Social Sciences to Enhance the Realism of Simulation for Complex Urban Environments

Stephen Kheh Chew Chai, Mohd. Faisal Bin Zainal Abiden,
Hui Min Ng, Shawn Thian, Sidney Tio, Serge Landry
and Antoine Fagette

Abstract In this paper we discuss how findings from social sciences research can be injected into a complex urban environment simulator in order to increase the level of realism of the simulated behaviors with respect to the local context. Our team, composed of engineers and social scientists, describe here our approach toward tackling complex simulation problems with embedded human factors. We present some of the results obtained from two different use cases. The rationale and methodology behind those results are further detailed and the limitations and future improvements required are highlighted. This paper shows how simulation should contribute to the improvement of the quality of life of every citizen in a Smart Nation.

1 Introduction

Building a livable and sustainable urban environment is a complex undertaking. Nowadays, architects and engineers turn to simulations to visualize and validate their designs prior to executing their plans. There are many tools for infrastructure and human crowd simulation on the market, some more mature than others, and most of them are engineering focused. However, making assumptions on infrastructure design using simulations based solely on either the engineer's or architect's point of view can lead to a complicated situation at best if not a hazardous one at worst. Traces of such situations can be found in the world around us, e.g. muddy shortcut beside intended sidewalks, bottlenecks at popular passageways, misplaced and/or undersized doors given the evolution of the flows of pedestrians following a

S.K.C. Chai (✉) · Mohd.F. Bin Zainal Abiden · H.M. Ng · S. Thian · S. Tio · S. Landry · A. Fagette
Thales Solutions Asia, 28 Changi North Rise, Singapore 498755, Singapore
e-mail: stephen.chai@asia.thalesgroup.com

A. Fagette
e-mail: antoine.fagette@asia.thalesgroup.com

© Springer International Publishing Switzerland 2016
M.-A. Cardin et al. (eds.), *Complex Systems Design & Management Asia*,
Advances in Intelligent Systems and Computing 426,
DOI 10.1007/978-3-319-29643-2_18

235

change of points of interest in the environment. On paper, the design may be sound but the critical human element is missed.

Wetmore in [1] argues that engineering might not have all the tools and techniques to get the job done. He suggests that engineers should borrow tools and techniques from social sciences in order to capture and measure the social implications of their decisions. In other terms, engineers and social scientists should collaborate to design people centric solutions.

Although simulations are enabled through technological means and run on mathematical models, modeling human behavior must inevitably involve the social sciences. In the context of urban simulation, we are particularly interested in human behavior of individuals, groups and crowds. The main disciplines under social sciences that deal with how people behave are Psychology and Sociology. While Psychology studies the intrinsic mechanisms of a human mind, Sociology explores the overall social behaviors of people in their environment. Therefore, our work attempts to connect the outputs of the social sciences as inputs for our engineered behavioral model.

In this paper, we first present in Sect. 2 a brief overview of the related work in the domain of human and crowd simulation. Section 3 then describes our methodology to adapt social sciences approaches to our own crowd simulator SE-Star and vice versa. In Sect. 4 we present some results obtained by our team composed of engineers and social scientists on two use cases. Finally, Sect. 5 discusses these results and the consequences they have for our future works in the domain of simulation, before concluding in Sect. 6.

2   Related Work

To create a high fidelity urban simulation, there are two critical factors to take into account, the human beings (also referred to as "the agents" in simulation) and the infrastructure. To take these two factors into account, it is necessary to model the social behaviors they exhibit.

For the first factor, modeling social behaviors down to the psychological aspect is best achieved through Agent Base Simulation (ABS). As described by Gilbert [2], "agent-based simulation offers the possibility of modeling individual heterogeneity, representing explicitly agents' decision rules, and situating agents in a geographical or another type of space". Davidsson [3] positions himself at the intersection of Computer Simulation, Agent-Based Computing and Social Sciences to define the Agent Based Social Simulation (ABSS) approach and how it can help cross-fertilization between the three areas. This approach is further described by Li et al. [4], where the authors identify the underlying social theories for ABSS from both individual agent and multi-agent system perspective.

Regarding the infrastructure, Farenc et al. [5] consider this factor with the notion of "Informed Environment". "Informed Environment" is helpful to create urban infrastructures that provide the virtual humans with knowledge of the environment, improving the human-infrastructure interactions using simple social behavior models. Digging deeper on the social behavior aspect, Musse and Thalmann [6] demonstrated the adoption of sociology to represent several behaviors and represent the visual output for visualization purposes. This approach focuses on the inter-personal interactions in the crowd and how a crowd, composed of individuals moving in various directions with common or different goals, avoids collisions. On individual behavior, Pelechano et al. [7] presented their approach of integrating a psychological model into a crowd simulation system while Chao and Li [8] worked on integrating sociology into the simulator.

In [9], Navarro et al. describe SE-Star, a crowd simulator that has been developed for the past ten years by Thales. "SE-Star is a microscopic multi-agent simulation, which is able to reproduce a large panel of human behaviors via the use of several completely independent processes". SE-Star, a variant of ABSS, is composed of three components, (i) the entities (or agents), (ii) the smart objects and (iii) the environment infrastructure. Each entity is unique and autonomous. A smart object represents an actual system (e.g. ticketing machine, ATM, ticket, signboard, etc.) and offers products and services to the entities in the simulated environment for them to realize actions and satisfy needs. The environment infrastructure is a 3D model of the environment to simulate.

In SE-Star, each entity's brain uses a motivational engine based on the Free-Flow Hierarchy approach that computes the physiological and psychological state of each agent, controlling its motivations and actions based on its traits, perceptions and internal variable as described on Fig. 1.

Fig. 1 SE-Star motivational tree

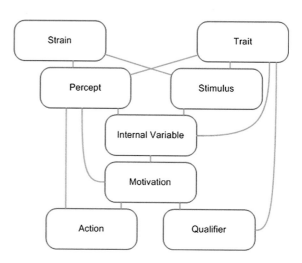

3   Approach

For this study, we chose SE-Star to conduct our simulations. The concepts of entities and smart objects that it encompasses, as well as the motivational engine that it implements for the brain of each agent make this simulator an ideal candidate for our study involving the inputs of social scientists.

3.1   Overview

To construct a realistic model for complex urban simulation, we divided the work into three phases, (i) input (Sect. 3.2), (ii) process (Sect. 3.3) and (iii) output (Sect. 3.4), as described in Fig. 2.

Following Wetmore's recommendations in [1], the team working on the subject includes members from both Engineering and Social Sciences backgrounds.

The first half of the team comprises of four engineers. The senior engineer within the four leads this half of the team to manage the 3D models, define the various simulation systems and help the social scientists with the implementation of the behavior models.

Four social scientists comprise the second half of the team. The psychology specialists model individual behaviors, and the sociology specialists look at social and group behaviors. In addition, a political scientist helps look at the policy implications on individuals, groups and infrastructure. Together, the social

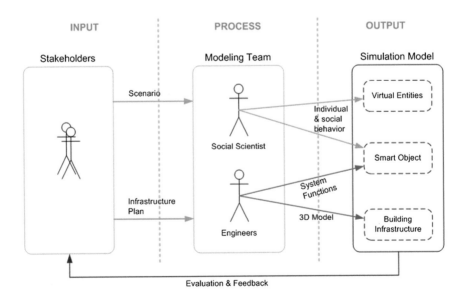

Fig. 2 Workflow for our approach

scientists are responsible for creating the new archetypes of people with distinct behaviors and unique characteristics for each scenario.

3.2   Input Phase

The initial phase of our approach is to identify the various stakeholders related to the environment to be simulated. They provide the context of the simulation as well as the various data and knowledge that they have gathered through experience and observation. Data can range from blueprints of the environment to statistics pertaining to the population to be modeled, while knowledge is more empirical in nature. The stakeholders identified provide feedback and validation on the simulation once the model is developed.

However, interaction with the stakeholders can also bias the behavioral models toward particular use cases, preventing a good generalization of these behavioral models. In this case a Systematic Framework, with reusable behavioral models based on established studies in Social Sciences, should be created. Such a framework not only reduces the effort when defining behaviors for new use cases by reusing models from contextually similar use cases, but also improves the quality of the simulation, unleashing the human-like innovative spirit in our simulation model.

In this Input phase, the engineering part of the team shall focus on the environmental infrastructure and system part of the problem while the social scientists shall gather from the stakeholders as much information as possible regarding the social behaviors and processes that they observe in their environment.

3.3   Process Phase

The Process phase is where the stakeholder's inputs are analyzed and transformed into a highly realistic simulation model. The environment to be simulated is virtualized into a 3D model, typically based on building or city plans provided by the stakeholders. The equipment and systems, known as smart objects, are analyzed and modeled. Smart objects' characteristics, the possible actions and reactions of the human beings populating the environment and observed by the stakeholders, are crucial details needed in the modeling process.

Furthermore, to complete the panel of data and knowledge provided by the stakeholders, psychological and sociological studies can be conducted on-site by the social scientists of the team. The result of this phase is one or more models enabling a realistic simulation of the environment.

The working experience between engineers and social scientists has to be managed delicately due to the inherent differences in terms of perspectives and skill sets between these two backgrounds. For instance, an engineer sees an Automated Teller Machine (ATM) as a machine dispensing money while a sociologist might see the correlations with age, socio-economic class, etc. Social scientists might find it hard to understand the computer language used to code the simulation models, whereas engineers from the same team could assist to overcome this weakness. Hence, the team's dynamics plays a significant role in getting the social scientists and engineers to understand and respect each other's points of view in generating a representative simulation model.

3.4   Output Phase

In the output phase, the team implements the model(s) derived during the previous phase into the three components of SE-Star. These are identified as the entities, the smart objects and the environment infrastructure. The links between the components are also established and simulations can be run to assess the validity of the models derived. Validation of the output is conducted together with the stakeholders. It includes the testing of various scenarios, the generation of 'What if…?' scenarios, and the tweaking of different variables the client is concerned about. The observations are then compiled into a report delivered to the client.

4   Results

In this Section, we present two case studies that have been successfully implemented using our approach and SE-Star. The first case study (Sect. 4.1) describes the different behaviors existing for the evacuation of a building when a fire drill is occurring versus during an actual fire. The second case study (Sect. 4.2) aims at describing the arrival of passengers in an airport and how they decide on their mean of transportation to leave the airport after collecting their luggage.

The work performed on these case studies has enabled us to test the limits and capabilities of SE-Star as well as to learn from these limitations and to suggest future improvements. More importantly, we were able to analyze the fit between social sciences data and a computer engineered framework, and therefore to confirm the possibility for these two domains to complement each other.

Hence, the objective of this paper is to create awareness on the benefit of coupling social sciences with complex systems. This statement may seem obvious but is in fact a challenging feat to implement. Few simulators available on the market incorporate social sciences insights and real data, and thus a gap remains between reality and simulation. It is hoped that the case studies mentioned in this paper, though descriptive, present some ways to overcome these gaps.

4.1 Case Study 1: Fire Evacuation

In this case study, we simulate a very familiar environment, namely our office building, and trigger a fire evacuation of this building in two different circumstances, (i) a fire drill and (ii) a real fire. The intention of this simulation is to contrast the situations and behaviors during the evacuation when it is a drill versus when it is a real emergency.

4.1.1 Input

Before going for an emergency situation (whether drill or real), the normal situation has to be modeled and simulated. For that, our social scientists have observed the behaviors of our fellow employees, their integration within the environment and their use of the various facilities on a typical working day. Our engineers came up with a 3D model of the building and the various facilities available. For the fire drill situation, our team relied on video surveillance footages as well as feedbacks from the team in charge of the security of the premises. On the other hand, for the real emergency evacuation it is quite impossible to observe how people act as there were no prior incidents. Therefore, two evacuation theories were used for that model, (i) the Emergent Norm Theory described by Kuligowski [11] and (ii) the Familiarity Model of Mawson [12].

4.1.2 Process

For the drill evacuation, based on what is observed, a simple broadcast of the PA system increases the virtual entities' motivation to evacuate. All staffs leave through the main entrance and gather at the assembly point outside of the premise.

Regarding the real emergency situation, we rely on (i) the Emergent Norm Theory [11] and (ii) the Familiarity Model [12]. To emulate these two theories, we modify smart objects to induce certain behaviors.

Emergent Norm Theory

This theory is based on norms. Norms generally are the rules and regulations that groups live by. Individuals go through day-to-day activities so often that they have become routines. However, disasters disrupt the normal routine of individuals. Faced with the unknown, individuals are required to make a concerted effort to create new meaning out of new and unfamiliar situations. This is done through the perception of environmental and social cues. The type of action chosen depends on the perceived characteristics of the threat. Hence, inhabitants will usually take some time to evacuate even after the fire alarm, or any other warning system, has activated.

In the simulation, the first stage is therefore to have the entities not evacuate immediately when the alarm sounds, but rather after the danger is confirmed by an individual. The fire will activate an invisible smart object near a workspace. The entity at this workspace will perceive the broadcast of the invisible smart object and he will have the motivation to search for the source of danger. He will venture out the door and perceive the fire. This will cause him to display the message "Fire!" The fire broadcast decreases the 'search motivation' and increases the entity 'inform motivation' which drives the entity to interact with 4 invisible smart objects in the office. The 'inform motivation' causes the 'inform qualifier' to activate. Others will perceive this qualifier and have their 'evacuation motivation' increased. Since the informing entity is going around the office interacting with invisible smart objects, it is as if he is trying to tell everyone of the danger because at every interaction he will display the "Fire!" message. At the final invisible smart object, he will have his 'inform need' reduced and 'evacuate need' increased, and thereafter will evacuate with the rest of the entities.

Familiarity Model

Rather than flight or flee, this model suggests that the typical response to danger is affiliation. People tend to turn to and protect loved ones even in the face of threats. Because of this, when people are forced to evacuate, they tend to do so in a group thereby maintaining proximity and close contact with familiar people.

During the simulation, as the entities perceive the 'inform qualifier', their 'evacuation need' increases. However, this is the first phase of the evacuation, named 'evacuation1 need', that will make the entities gather at the corridor. There, two invisible smart objects give off broadcasts. One of these invisible smart objects reduces 'evacuation1 need' while the other will increase 'evacuation2 need'. This takes a few moments so it is as if they are stopping to wait for their peers. Subsequently, 'evacuation2 need' makes them gather at the assembly point.

4.1.3 Output

The fire evacuation simulation was benchmarked against video surveillance footages and verification from staff members. In a simulation without social behavior, the entities will start evacuating as soon as the fire alarm is triggered and they will strictly follow the predefined process flow. The entities enriched with social sciences cues exhibit behaviors as suggested in [11, 12].

The results demonstrate that social behavior theories can be translated into our simulator to simulate drills of emergency situations and real emergencies. Such models should be of interest to design drills that are more realistic but also to assess the quality of the results of the drill versus how people would have performed should the situation have been a real emergency.

4.2 Case Study 2: Transportation Choice

This case study models a busy airport where arriving passengers are faced with a decision to take connecting transport to the city. We intend to create the generic model passengers archetype and simulation model. This model can then be used to study the impact of environmental changes on passengers' behaviors, e.g. how the closure of certain areas due to renovations is affecting passengers' decision to take a particular mode of transport.

4.2.1 Input

Our engineers started by constructing the 3D model of an airport and modeled the smart objects in this airport arrival hall. With limited stakeholders input, the social sciences team worked on sourcing for data related to people's perceptions of different transport systems in Singapore. The team decided to use publicly accessible data such as [13–16]. These data enable us to understand peoples' considerations when making a decision on their choice of transport and how different groups of people evaluate the different modes of transport differently.

4.2.2 Process

The process phase went through two steps, to (i) determine the various personality traits and social characteristics each entity will exhibit influencing its final decision and (ii) identify the archetypes of entities and how each predetermined traits and characteristics will vary in these archetypes.

Personality Traits and Social Characteristics
Human behaviors are often quite complex in the sense that they have many determinants. Some of these determinants might involve acquired and innate physiological factors (learned or inherited behavioral predispositions), or they could include environmental causes and situational exigencies. Thus, the first step to predict the behavior of a group or individual is to identify and learn its determinants. In the context of travelers deciding on the mode of transport to take from an airport, what influences the decision to take a taxi, the MRT, or a bus is very much related to one's consumption pattern, shaped by one's socio-economic class, thriftiness and association with materialistic values.

The diagram presented on Fig. 3 is a simplified version of SE-Star's motivational tree developed by our team specifically for this use case. It illustrates the decision processes and the inter-connectivity of various factors in the entities' brain, starting from traits and strains at the top, leading to their final action to take a taxi (in this example) at the bottom.

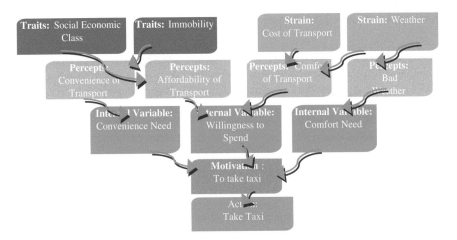

Fig. 3 SE-Star's motivational tree leading to the action to take a taxi

	Locals	Tourists	Business travelers
Socio-economic class	Varies	Varies	Varies
Convenience need	High	High	High
Willingness to spend	Low	High	High
Number of luggage	High	Low	Low

Table 1 The different profiles of entities and the variations of some traits and strains

Profiles of Entities Archetypes

The simulation features three main archetypes of passengers, (i) Locals, (ii) Tourists and (iii) Business Travelers that depict the crowd in the airport arrival hall on a typical day. The archetypes are based on the three most common passengers defined by the stakeholders. Sub-archetypes variants can be derived based on the main archetypes. The Table 1 presents a generalization of different profiles of entities coupled with an excerpt of the traits and strains.

4.2.3 Output

We have modeled a generic airport terminal with localized social behaviors. We observed expected and unexpected behaviors that were not predetermined such as the swarming behavior toward the luggage carousel when the first pieces of luggage appear, lost tourists looking for information due to the lack of local knowledge and the formation of queues at taxi stands and bus stops.

During our fieldwork we also observed behaviors that are not reproduced by our simulator such as impatient passengers who tend to cut queues, families who tend to cluster around a point in the queue, couples who tend to be side by side in a queue.

All these group characteristics and behaviors are still lacking in our current model, an area identified for our future research.

Using the same behavioral models, it will be possible to test at a site or township level the transportation capacity in various scenarios such as passenger increase (*e.g.* due to an event), infrastructure works or a breakdown. This approach could be scaled up to the level of the entire city transportation system.

5 Future Work

Having involved social sciences in our simulation modeling, we have uncovered valuable insights into the strength and weaknesses of our simulator SE-Star. We shall look into improving Group Behavior modeling. The notion of Group Brain and Group Objective where two or more individuals having the same goals share a "common brain" and inter-entities communication could be further explored. At a macro level, simulations should consider social structures and implication of culture influencing the desires, beliefs and values.

Regarding social sciences itself as contributor to SE-Star, we find out that much as psychology easily finds its place into the study, sociology's angle of attack to contribute appears less obvious. This direction of study needs to be further researched, through further field researches, modification of the simulation engine or how the smart objects are apprehended by the virtual entities.

6 Conclusion

In this paper, we presented our integrated team composed of both engineers and social scientists and its approach to enhance the realism of a simulator for complex urban environments. We also presented some results of simulations produced by that team, using our crowd simulator SE-Star.

The newly formed multi-disciplinary team has proven to be very effective in tackling complex problems, introducing new social behaviors into our virtual entities, and thus enhancing the simulation fidelity. Members were able to contribute in their respective domains of expertise, helping to analyze and integrate the stakeholders' requirements into those new models. From our case studies we uncovered that some social sciences theories can indeed be implemented in SE-Star, in particular those pertaining to psychology, while some others, related to sociology, would require some modifications of our engine.

For future work we intend to further improve group behavior and inter-entity communication in SE-Star. We shall further take into consideration the impact of social sciences to improve our current approach, focusing in particular on strengthening the contributions of psychology and improving the weight of sociology.

We believe the approach we have described could be used as a reference to build more "humanistic" simulation models. Those models should enable engineers to simulate the social implications of their work, thus minimize the mismatch of expectation between the intended design and the desires of the citizens, effectively contributing to the improvement of the quality of life.

Acknowledgment We would like to express our appreciation to team members from National University of Singapore (NUS) and Nanyang Technological University (NTU) who have contributed to this project during their summer internship 2015, with special thanks to Choo Xin Wei and Lai Shi Min for their contribution.

References

1. Wetmore, J.M.: The value of the social sciences for maximizing the public benefits of engineering. Bridge **42**(3), 40–45 (2012)
2. Gilbert, G.N.: Agent-based models (No. 153). Sage (2008)
3. Davidsson, P.: Agent based social simulation: a computer science view. J. Artif. Soc. Soc. Simul. **5**(1) (2002)
4. Li, X., Mao, W., Zeng, D., Wang, F.Y.: Agent-based social simulation and modeling in social computing. In: Intelligence and Security Informatics, pp. 401–412 (2008)
5. Farenc, N., Boulic, R., Thalmann, D.: An informed environment dedicated to the simulation of virtual humans in urban context. In: Computer Graphics Forum, vol. 18, issue 3, pp. 309–318 (1999)
6. Musse, S. R., & Thalmann, D.: A Model of Human Crowd Behavior: Group Inter-relationship and Collision Detection Analysis, pp. 39–51. Springer Vienna (1997)
7. Pelechano, N., O'Brien, K., Silverman, B., Badler, N.: Crowd simulation incorporating agent psychological models, roles and communication. Pennsylvania University, Philadelphia Center for Human Modeling and Simulation (2005)
8. Chao, W.M., Li, T.Y.: Simulation of social behaviors in virtual crowd. In: Computer Animation and Social Agents (2010)
9. Navarro, L., Flacher, F., Meyer, C.: SE-Star: A large-scale human behavior simulation for planning, decision-making and training. In: Proceedings of the 2015 International Conference on Autonomous Agents and Multiagent Systems, pp. 1939–1940 (2015)
10. McCowan, I., Bengio, S., Gatica-Perez, D., Lathoud, G., Monay, F., Moore, D., Wellner, P., Bourlard, H.: Modeling human interaction in meetings. In: Proceedings of the IEEE International Conference on Acoustics, Speech, and Signal Processing, vol. 4, pp. IV-748 (2003)
11. Kuligowski, E.: Predicting human behavior during fires. Fire Technol. **49**(1), 101–120 (2013)
12. Mawson, A.R.: Understanding mass panic and other collective responses to threat and disaster. Psychiatry **68**(2), 95–113 (2005)
13. Land Transport Authority of Singapore (LTA): Household Interview Travel Survey (HITS) 2012: Public transport mode share rises to 63 % (2013). http://www.lta.gov.sg/
14. Land Transport Authority of Singapore (LTA). Taxi Customer Satisfaction Survey (TCSS) 2013 (2014). http://www.lta.gov.sg/
15. Institute of Service Excellence (ISE): Customer Satisfaction Index of Singapore (CSISG) 2013 (2013). http://ises.smu.edu.sg/
16. Institute of Service Excellence (ISE): Customer Satisfaction Index of Singapore (CSISG) 2014 (2014). http://ises.smu.edu.sg/

Inferring Activities and Optimal Trips: Lessons From Singapore's National Science Experiment

Barnabé Monnot, Erik Wilhelm, Georgios Piliouras, Yuren Zhou,
Daniel Dahlmeier, Hai Yun Lu and Wang Jin

Abstract The following paper presents three novel and efficient algorithms to tackle pressing questions asked by city planners as well as policy makers: Where are people starting and ending their trips? Which activities are people traveling to/from? Are they taking the most efficient route? In order to capture large-scale travel data, a novel sensor was developed by the Singapore University of Technology and Design together with industrial partners. Using computationally simple and scalable algorithms, we are able to understand the large amounts of data collected by the sensors and shed light on the three questions above.

Keywords Urban data · Large-scale experiment · Sensor data · Optimal routing · Data visualization

B. Monnot (✉) · G. Piliouras
Engineering Systems and Design, Singapore University of Technology
and Design, 8 Somapah Road, Singapore 487372, Singapore
e-mail: monnot_barnabe@mymail.sutd.edu.sg

G. Piliouras
e-mail: georgios.piliouras@gmail.com

E. Wilhelm · Y. Zhou
Engineering Product Development, Singapore University of Technology
and Design, 8 Somapah Road, Singapore 487372, Singapore
e-mail: erikwilhelm@sutd.edu.sg

Y. Zhou
e-mail: yuren_zhou@mymail.sutd.edu.sg

D. Dahlmeier · H.Y. Lu · W. Jin
SAP Singapore, 1 Create Way, #14-01/02, Singapore 138602, Singapore
e-mail: d.dahlmeier@sap.com

H.Y. Lu
e-mail: hai.yun.lu@sap.com

W. Jin
e-mail: crystal.wangjin@gmail.com

© Springer International Publishing Switzerland 2016
M.-A. Cardin et al. (eds.), *Complex Systems Design & Management Asia*,
Advances in Intelligent Systems and Computing 426,
DOI 10.1007/978-3-319-29643-2_19

247

1 Introduction

The study of movement and travel patterns for residents of mega-cities is an important and growing field. The results of this field of study are consumed by city planners, transport planners, and policy makers who must manage systems which are steadily growing in complexity as more people move into urban centers.

We look in this paper at three of the most common questions asked by these agents:

1. Where are people starting and ending their trips?
2. Which activities are people traveling to/from?
3. Are they taking the most efficient route?

To tackle these questions, we make use of data gathered by a novel sensor developed by the Singapore University of Technology and Design together with industrial partners. The main contributions described in this paper are three novel and efficient algorithms for accurately answering the above questions for very large data sets. Building origin/destination matrices have often previously been effectively accomplished with GIS methods based on GSM (cellular phone) data [1] but this approach is limited by the accuracy of the GSM localization data which is often imprecise. The algorithms presented here are computationally simple and therefore scalable. Using embedded systems for activity identification has previously been complex, memory intensive, and energy intensive; our algorithms reduce the task's complexity and accurately identify a small set of activities using little memory and energy [2]. Previous studies of route identification have the luxury of more plentiful geolocation data [3], whereas our algorithms are capable of studying route efficiency using sparse Wi-Fi geolocation data. The contributions described in this work should be useful for research and policy teams using smart-phone applications, GPS loggers, and 'Internet of Things' sensors alike to collect and study human mobility in urban centers.

2 Singapore's National Science Experiment

As part of a major Smart Nation initiative, a sensor (called the SENSg) was custom-designed by the Singapore University of Technology and Design (SUTD) together with the Delta Electronics Industrial Automation Business Group. The goal of the sensor and accompanying server infrastructure designed by the SUTD together with the A*Star IHPC group is to inspire and thereby motivate Singapore's students to pursue science and engineering related fields. The sensors gather temperature, relative humidity, light level, sound pressure level, atmospheric pressure, 9-degree of freedom motion data. Additionally, they possess a Wi-Fi radio which serves the dual purpose of scanning for MAC addresses which are used to localize the sensor nodes as well as to move sensor data to a back-end server. The sensors

Fig. 1 The third generation SENSg device (*left*) which is worn by students on a lanyard (*center*) which allows them to explore their personal travel and environmental data (*right*)

will be deployed on a large scale from September until November 2015. The sensor shown in Fig. 1 is designed to acquire students step count, determine whether a device is located indoors or outdoors, as well as automatically identify travel mode (walking, riding a bus, riding a train etc.).

In order to validate the performance of the sensors before they are deployed, a series of three pilots were arranged at various primary and secondary schools in Singapore from June through July 2015. In total, 300 sensors were distributed and over 4.5 million unique sensor measurements were made. From this set, over 400 trips were able to be determined using the methods outlined in the remainder of this paper. The data shown in Fig. 2 are indicative of the types of data which can be acquired using the SENSg sensors. The localization of measurements is accomplished using the Skyhook API which returns geographical coordinates when it is sent a list of MAC addresses from Wi-Fi access points which were scanned by the device.

Fig. 2 Density of Wi-Fi network accesses recorded by the first phase of sensor deployment

3   State of the Art

3.1   Trip Inference

Inferring trips from geographical data has been tackled before in the literature. The first question is how to divide an agent's sequence of locations into individual trips?

Two approaches have been employed: active or passive trip reporting. In the active method, agents in the experimenting are tasked with providing the researchers a detailed account of their trips, including the nature of the trip. This allows the recuperation of verifiable information but has one major shortcoming: agents will be likely to under-report their trips, due to the cumbersome nature of logging their activities every time (a shortcoming noted by [4]).

Passive trip reporting offers a promising alternative that also scales better: only raw data such as the agent's latitude and longitude at a given moment is collected periodically; it is then the task of the researchers to process the data to break the sequence of positions into trips. The burden of verifying the agents' logs and building a database from it is avoided, which makes the method more suitable for large data sets.

Several studies have made use of passive trip reporting. In [5], the agents' cars are tracked. A trip is then defined as a sequence starting from the powering on of the car and ending when the contact is off (with additional criteria to filter out smaller trips or include stops during which the car is left on).

Schüssler and Axhausen [6] tracks the person's location instead of their car, which multiplies the post-processing difficulties due to the continuous nature of human movements. The researchers applied a smoothing method on the geographical data points to overcome the deficiencies of GPS-based data collection (listed in [7]). Trip inference is made by considering bundles of data points, i.e. logs that are geographically close to each other, and defining thresholds that discriminate between a stop within a trip or the endpoint of one. These thresholds can be based on the agent's speed (computed from the difference between two successive data points) or the time spent in the same bundle of points.

One of the most vexing problems facing researchers studying large-scale mobility patterns is the unwillingness of participants to spend their smart-phone battery energy for collecting location data from power-hungry GPS services. Several research groups (including this one) have proposed down-sampling location estimates as a solution [8, 9]. An alternative approach was taken in the design of the sensor system for the National Science Experiment which was described in detail in Sect. 2. For acquiring participant locations, the MAC addresses of surrounding Wi-Fi hotspots are scanned and recorded. This has been shown to be accurate to within 20–30 m and tends to be more effective while walking than driving [10].

3.2 Activity Matching

Trip inference and activity matching are two closely related problems. Once we have found out which trips the agent has taken, we want to know the nature of these trips. Again, we can ask the agents directly to provide information about the trips, but we expose ourselves to the same deficiencies noted before. We then seek an algorithmic way of assigning activity to a trip.

Stopher and FitzGerald [11] uses GPS data with additional information such as home and workplaces of the tracked agents to assign purpose to the trip. For moves that do not possess either or both as endpoints, they rely on land use data, i.e. geographical data giving the function of a particular piece of land (e.g., educational, hospital, road, park etc.) [12] uses both GIS land data and agent validation through a web application to confirm the activity.

In [13] theoretical frameworks are defined that allow for capturing the spatial and temporal phenomena of daily travel. Finally, a recent survey [14] focuses on the emerging trend of identifying semantically rich trajectories, instead of merely raw movement data, as the main object of interest in mobility studies. It describes a host of different techniques developed for addressing diverse issues such as creating trajectories from movement data, overlaying semantic information to trajectories and using data mining to extract higher level understanding of the characteristics of the trajectories.

4 Data Analysis

4.1 From Raw Data to Trip Recognition

4.1.1 Inferring Trips From Wi-Fi Data

The challenge of identifying travel mode from sparse Wi-Fi localization data was previously treated by this research group using a k-means clustering approach to identify home and school locations [15]. The main issue with applying clustering algorithms to this problem is that there are often substantially more points available for public spaces than for home locations leading choose bus interchanges and malls being identified as places where more time is spent, as opposed to true home locations. This paper takes an approach which considers the velocity of the person carrying the device as the primary indicator of the start or end of a trip. The dwell-time based algorithm presented here identifies static points in the localized coordinates by calculating an approximation of mean device velocity. It is less prone to mis-identification of home/school clusters because it filters for locations recorded in the late evening and during prime school hours to attempt to ensure that the POI's are correctly identified based on the probability that a point recorded at a given time corresponds to a student's location. The algorithm has the following steps:

Algorithm 1 Determining trips from dwell time

Require: Latitude, longitude, timestamps, stopped threshold S_{thresh} e.g. 0.1m/s, dwell time threshold D_{thresh} e.g. 240s, student at home e.g. 22h-5h, student at school e.g. 9h-12h

Round latitude and longitude to 4 sig. dig. to ensure 10m maximum accuracy

Load GIS data

for point in lat/lon data **do**

    Calculate distance between each point using Haversine

    Calculate velocity between each point using $\dot{x} = \Delta x / \Delta t$

    Smooth velocity data using moving average filter (e.g. 3 windows)

    Count number of stopped points N_{points}

    if $\dot{x} < S_{thresh}$ **and** $N_{points} > D_{thresh}$ **then**

        Store lat/lon in list **POI**

Find unique *POI* in list **POI**

Check if unique *POI* occurred during home/school time in ts

Assign $POI = POI_{home}$ and $POI = POI_{school}$ based on maximum occurrences of unique **POI** during home/school time

Check that distance between POI_{home} and POI_{school} is >1000m

Calculate distances between remaining unique **POI**

Source code for this algorithm in python and MATLAB can be downloaded here: SUTD Code Repository. Public sample data sets beyond the sample data contained in the repository may be requested.

When running the algorithm on a data set from a pilot test consisting of roughly 100 sensors deployed over a period of 2 days, Fig. 3 shows how the dwell-time algorithm is able to successfully identify approximately 50 % of the trips which were taken from home to school.

4.2 Matching POI's with Activities

From the previous trip analysis and temporal data, we can infer the students' homes and schools locations. The trips however may start or stop at different endpoints, such as a restaurant where students go to eat, shopping malls or commercial areas where they gather or parks and sports installation for recreational activities. We would like to be able to tell which of these activities, if any, is being undertaken by the student.

Fig. 3 Trips from home to school and back calculated using velocity dwell time (*red* home, *yellow* school)

To answer this, we obtain information from three different channels:

- Land use data
- Google Maps Places API data
- Feedback and validation from the students

The first part is to precisely identify the different POIs that the student is visiting. We define as POI an endpoint to a trip. Since we collect many of them in the first place, it is useful to identify those that are very close to one another and define them as one POI. This process reduces the redundancies.

If the trip analysis algorithm was successful, we have already identified the home as well as the school of the student. If not, it is possible to use a cruder method of counting the logs around the POIs, and decide that the one with the most logs will be the home and the one with the second most number of logs will be the school. For the other POIs, we proceed in the following way. Weights are assigned to each type of activity (food, commercial, recreation) and updated using the three channels of information. At the end of the process, we assign to the POI the activity that we find the most weighted.

The land use data gives us polygon representations of parks and sports installations in Singapore, which we can use to decide if a given POI is inside these polygons or not. This is a very strong clue to decide that the student is engaging in a recreational activity. However, Singapore is a very dense place that has many food outlets or stores even in parks (or the reverse: the Kallang mall has a climbing wall in the middle of the other shops). We therefore assign a consequent weight to the "recreation" activity if we find out that the POI is inside one of these parks and sports polygons yet not consequent enough to decide once and for all that the POI is recreational.

The Google Places API comes in handy to get more information about the places around a particular POI. We request the API a list of nearby places that are of a certain type (for our analysis, food, shopping and sports related places). The request returns this list with the location (latitude and longitude) and types of the places. We then update our weights once again with the results of this request. If the request has returned an item of, say, "shopping" type, we add some weight to the shopping activity, *inversely proportional to the distance to the POI*, so that closer places add more weight.

With these two channels we can already discriminate between the three activities. It follows a best guess approach that gets more confident with the number of results returned by the Google API, a technique that does not appear in the previous literature. We are able to assign the POIs with an activity and understand better the nature of the trips taken by the students. We hope a smaller amount of activities are classified as "Others" as our database of POIs and our student input grow.

To drive the point home we finally give the student the ability to cross-check the results of our analysis in the most painless way possible. Previous studies have underlined the difficulty of collecting the data from the participants in an active way, where they would be asked to provide themselves information about their trips. This usually leads to an under-reporting of the trips taken by the agent because of the cumbersome logging. By integrating in our interactive web-application a system that asks for the most uncertain POIs, say, "Is this where you get some food? Yes or No" with a visual map and position of the POI, we are able to correct deficiencies of the algorithm and even build a database of labeled examples that we can reuse in further studies. This is closer in spirit to the approach followed in [12].

We give in Algorithm 2 a pseudocode explaining how to assign activities. Source code for this algorithm in Python can be downloaded here: SUTD Code Repository.

Algorithm 2 Assigning activity to POIs

Require: Latitude, longitude of the nodes; GIS data of recreational surfaces (sports/parks).
  Load node data
  Load GIS data
  for node in node data **do**
    Get trips from node
    Get POIs from trips
    Eliminate redundant POIs
    for POI in obtained POIs **do**
      if POI was assigned home or school **then**
        Assign home or school to POI
      else
        if POI is inside recreational land **then**
          Assign weight to recreational activity
        Load data from Google Places API
        for result in Places API response **do**
          Assign weight to activity linked to result
        if Weights have been collected **then**
          Decide POI type from most weighted activity
        else
          Assign others to POI
    Build origin-destination matrix for node from assigned activities

---

5 Engaging in Activities

5.1 Activity Classification

The Singapore National Science Experiment will give access to unprecedented amounts of data collected over several days. The algorithms described before will process a large amount of it, yet they could often use a bit of guidance from the human to corroborate the results and gain valuable insights from them. The addition of a visual component to understand large data sets has been recognized in such fields as fraud detection (see [16] and the older but influential [17] for two different accounts of this). We therefore seek to empower our users, which comprise both the researchers attached to the project as well as the students carrying the sensors, by giving them access to visual representation of the data from different angles.

Earlier, we have touched upon the cross-validation made by the students participating in the experiment: after our algorithms have assigned activities to the POIs, we confirm these activities from simple questions answered by the students. This helps us build a database of the POIs which we can then use to ask ourselves the questions: How often do students take trips from school to sports? From a food place to home? Understanding these questions gives us a crucial insight in the behavior of our agents.

One way to represent this information is by using a chord diagram. The underlying data structure is a matrix which we call *origin-destination matrix*. The origin-destination matrix T is a square matrix of size the number of different activities we are able to differentiate (in our study, there are six: Home, School, Food, Commercial, Recreation and Others). $T_{i,j}$ captures the number of trips made from activity i to activity j. We compute for each student s a origin-destination matrix T^s and sum it all up to get the total trip information T of all our agents.

We give below an example of chord diagram obtained from the total origin-destination matrix of one pilot session. The chord diagram is generated in Javascript using the D3 library [18]:

Around the outer ring of Fig. 4 we have labels giving the different activities. This ring is divided in six unequal portions, each one with a different color corresponding to a particular activity (e.g. light green for Recreation). The size of the portion is proportional to the number of trip start points recognized as belonging to the corresponding activity, which for activity i is the number $\sum_j T_{i,j}$.

From the outer ring we then have multiple chords connecting one portion to another. They hold the information given by our $T_{i,j}$ coefficients. Since the diagram can quickly become cluttered, we give the user the ability to fade out parts of it to make it more readable. This is shown in Fig. 5.

In the example above, we hovered over the Commercial activity. Look at the chord joining this activity to the Others portion, of dark blue tint. The size of the

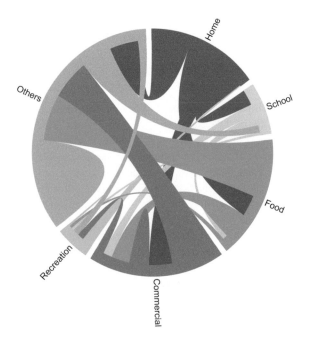

Fig. 4 Chord diagram from total origin-destination matrix

Fig. 5 Faded chord diagram
from total origin-destination
matrix

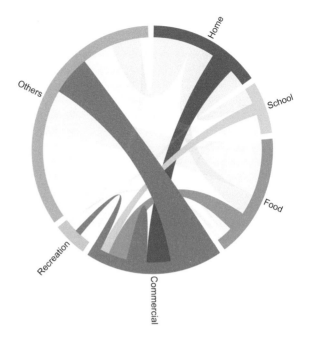

area covered by the chord on the Commercial portion is given by the number $T_{commercial,others}$, i.e. the number of trips going from a Commercial activity to one filed under "Others". Reciprocally, the area covered by the end of the chord, in the "Others" portion of the ring, has a size corresponding to $T_{others,commercial}$.

The chord's color is chosen to be the color of the activity that has the largest of the two chord's endpoints attached to it. In the Commercial to Others chord, we can see that the Commercial activity has the largest endpoint, i.e. $T_{commercial,others} > T_{others,commercial}$, so we pick dark blue for the chord's color.

Finally, for "power users" such as the researchers attached to the project, we build a complete web-app that gives access to each particular sensor data, completed by a map of Singapore with a log of the data points and the POIs (Fig. 6). To access the underlying data, we supplement the diagram with more information in a readable way, as shown below. As a portion is hovered over, we update the trips details on the right panel under the chord diagram.

On the far left is given the list of all the recorded users. Clicking on one plots the logs collected by the sensor on the map (in light red) and the POIs (filled with their assigned activity color and a black stroke). Hovering over a particular activity on the right-side chord diagram fades out the POIs not related to that activity to make the ones that are more prominent. The preliminary HTML/Javascript file used to create the visualization is given in our SUTD Code Repository.

We believe that this visual representation of the data allows a very quick understanding of the raw data logs collected from the students. A modified version of this application will be presented to the students for the "second opinion" that they are able to provide.

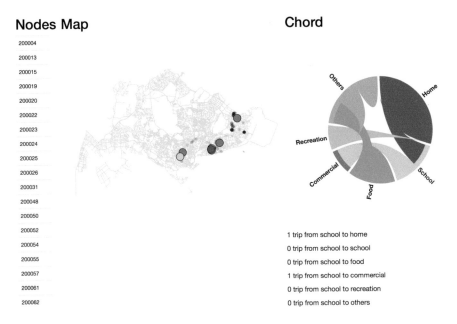

Fig. 6 Full application

5.2 *Preliminary Results*

Being able to precisely match activities with frequent locations visited by the students opens the door to many interesting results. The first and most evident one is to assess the quality of public policy measures. Singapore is known to be one of the greenest capitals in South East Asia (the National Parks Board motto even states "Make Singapore our garden") and to provide a considerable amount of sports installations around the city. Collecting data about trips and activities from middle school students allows a quick evaluation of just how much these public facilities are being used. They also provide more back-story to it: do residents have to make long trips to get to a recreational activity, and which places are the least equipped? Supplemented with mode identification to learn the way of transportation chosen by the students to reach a certain location, we can infer more information on the accessibility of these installations.

Singapore also possesses an incredibly diversified food culture, ranging from very local "hawker centres" (inexpensive food courts) to higher end mall restaurants. We can again query the sensor data to understand the behavior of Singapore students when faced with so many different food options.

So far the experiment has been run three times with a little over 300 students, each time in one school. While it is difficult to precisely answer the questions above, it is still fruitful to look at the collected data. We built up the total origin-destination matrix from the third pilot run, which is given in Fig. 4. It can already be seen that trips involving recreational activities (in our analysis, engaging in a sport or being in a park) are fewer than those with a commercial or food-related purpose.

An interesting development will be to see how precise our analysis was when feedback is received from the students, which has not been done yet. It will be a further challenge to fine-tune our algorithm given these responses from the students.

6 Choosing the Best Route

6.1 Visualizing Individual Trips

We now turn our attention to the itineraries picked by the students to join two locations. We want to know whether their choice of route is optimal, i.e. if they went by the fastest possible way from one POI to another. Since these POIs are highly familiar locations for the students (such as their home, school…), it is expected that they would have already optimized these moves as much as they can. If discrepancies are to be found, we then need to look into other data which the sensors can give us: did they pick the coldest route? The most shaded one?

To compare their trips with some real data, we use the Google Directions API to return the best possible route between two endpoints, either by car, by public transports or on foot.

The trips are then analysed using two different tools. We extend our previous application to plot students' trips compared to the Google Directions results. Some results can be seen in Figs. 7 and 8. We also develop some algorithmic procedures to remove trips that do not fit their Directions result (possibly because of a mis-classified trip or because of the noise in the sensor data) and compute some statistics on the remaining ones.

Fig. 7 Student's path (*orange dots*) and directions API result (*blue line*) for cars

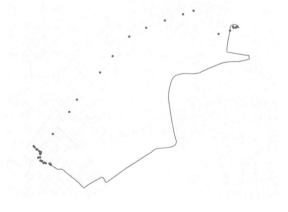

Fig. 8 Student's path
(*orange dots*) and directions
API result (*blue line*) for
public transports

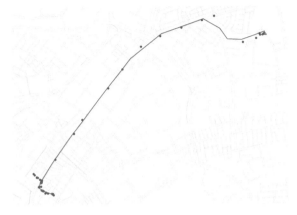

More specifically, we remove from our analysis trips that present the following characteristics:

- Trips that are too long (over an hour): these trips mostly appear due to noise in the location points that extend the trip beyond its true end.
- Trips that are not large enough (under 750 m): too short trips often get logged.
- Trips without enough data points (under 10): computing an accurate measure of the distance covered by the student is difficult if the data is too sparse.
- Trips that present too many gaps: again, we cannot have a precise distance measurement if logs disappear for too long. We use a metric based on the mean distance between two successive data points to remove these trips.

These criteria give us a coarser set of trips than that obtained for the previous activity detection analysis. One reason for that is that with no a priori on the nature of the trip, we could imagine a student going from his house to a nearby grocery store, which would be interesting to look at from the point of view of activity detection. However the best path analysis does not offer as much information for a very short trip.

In the Fig. 7 below, we plot one of the students' trip (orange dots) and compare it with the result from Google Directions (blue line). The figure is obtained by querying Google's API for the fastest way using a motorized vehicle. However, if we switch that parameter to return the fastest way by public transport, we obtain for the same trip a different picture that explains our data much better (see Fig. 8).

6.2 Optimal Routes: Some Results

The result presented in Figs. 7 and 8 has some interesting implications. The first and most obvious one is the possibility of using this kind of analysis for mode detection, another very active research question. Mode detection is tackled by our team using different methods, of which more information can be found in [15].

We can also apply this method to understand just how much more distance we have to make when we switch from a private mode of transportation to a public one. For example, we can make an informed public policy decision by deciding to add a new line of bus transportation in the places where taking one's private car is so much easier than using public transports.

This last part is better approached from an algorithmic perspective. The third pilot data set contains logs from over a hundred students, from which we identify about 300 trips. From these trips, we selected 51 of them which confirm to the set of criteria exposed in Sect. 6.1. We can then compare the distance covered by the students and the duration of their trips to the ones returned by the Directions API.

Figure 9 below charts the difference between the duration of a student's trip and the result returned by the Directions API. As we do not possess yet the mode of transportation used by the student, we select as a best guess the closest duration that the API returns to the student's. We observe in that case that most trips fall into a −20 to 20 % band of difference between the two durations. A finer analysis will be developed when the mode becomes available as an exterior measure computed from the data points.

We also count on the greater number of data points that will be collected in future runs of the experiments to erase local effects that we may be observing here, due to the fact that all students go to the same school. It will be interesting to see how the analysis will translate to more central or less dense parts of the city.

Finally, optimality also informs us on the congestion of the network for motorized trips. Maybe the user diverted from the fastest route as returned by the

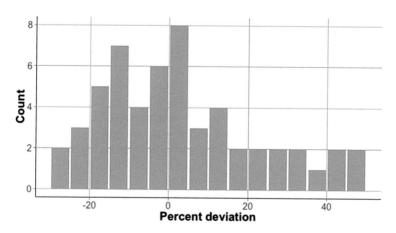

Fig. 9 Histogram showing the percentage of difference between the student's trip duration and the directions API's result

API (returned without using any information about the traffic at that time) because of the presence of many other users around him. We can then relate the making of the decision by the agent to the literature on congestion games with private information ([19] for the definition and models of congestion games). On the pedestrian level, the "cost" of taking a particular route could include such variables as the temperature (choosing to go to an air-conditioned indoors over a high temperature path outdoors), the light or noise levels (walking through a park instead of a residential area). These variables can be read in the data collected by the sensors and will be made use of in future experiments.

7   Future Work and Emerging Challenges

Singapore's National Science Experiment (NSE) opens up interesting possibilities about understanding the behavioral and social landscape of young Singaporeans. The project will involve more than 250,000 Singapore students over its 3 year lifespan who will be assigned easy-to-use devices that employ a number of sensors (location, motion, temperature, humidity as well as noise levels). Mining this data raises a number of interesting technical and analytical challenges that we briefly review next.

7.1   Big Data Algorithmic Techniques

Scaling up to high volumes of users creates new algorithmic challenges especially as each device broadcasts streams of data to a central server. Advanced algorithmic techniques such as streaming algorithms will be useful to deal with this massive data accumulation. These are types of algorithms that have limited memory available to them (significantly less than the input size) and moreover permit limited processing time per individual item but nevertheless can compute useful statistics or "sketches" of the data [20, 21]. More generally, sublinear time algorithms focus on computing functions of a target dataset while reading only a minuscule fraction of the whole input. A lot of recent work in computer science has focused on developing such techniques for a variety of algorithmic questions (see [22] and references therein). A particular notion of approximation which has been used to provide sublinear time algorithms for several problems is called property testing. These algorithms are used to decide if some object (e.g. a graph) has a "global" property (e.g., a graph is bipartite), or is far from having this property (e.g. the graph cannot be made bipartite even after removing a constant fraction of edges), while at the same time only using a small number of queries.

7.2 From Data to Stories: Understanding Real Life Social Dynamics and Networks

The ability to examine these diverse streams of sensor data gives rise to another interesting challenge. How can these data be woven together to create meaningful, high level semantic summaries about the true nature of the underlying social behavior? These data streams allow us to study two different types of correlations. The easier one is to combine the measurements of different sensors on the same device. This enables us to create more detailed and accurate descriptions about the individual behavior of each student. Critically, a different type of correlation is identifying correlations between users. In doing so, one can identify different types of groups that an individual belongs to along with the different group characteristics (e.g. size, dynamics, stability, etc.). This is a significantly more computational intensive task than the ones arising from focusing on individual users and applying sophisticated algorithmic techniques such as sublinear time algorithms will be necessary in order to deal with the emergent amount of data.

7.3 Privacy

Accessing detailed information about individuals' behavior naturally raises privacy concerns. Such privacy issues have been examined previously in the literature of semantic trajectories modeling and analysis [14]. As we move forward towards a full fledged realization of the vision of smart cities, deployed sensors will become increasingly present in the urban landscape. Any such progressive deployment should also be accompanied with the careful development of privacy preserving techniques that allow both for useful aggregation of data while minimizing the privacy loss for the individuals.

Acknowledgments This work was supported by the Singaporean National Research Foundation (NRF) and the SUTD International Design Center (IDC). Production of the sensors was possible due to strong support from Delta Electronics DRC, IABG, and Taoyuan Factory 2.

References

1. Caceres, N., Wideberg, J.P., Benitez, F.G.: Deriving origin destination data from a mobile phone network. Intell. Transp. Syst. IET **1**(1), 15–26 (2007)
2. Choudhury, T., Consolvo, S., Harrison, B., Hightower, J., Lamarca, A., LeGrand, L., Rahimi, A., Adam Rea, G., Bordello, B.H., et al.: The mobile sensing platform: an embedded activity recognition system. IEEE Pervasive Comput. **7**(2), 32–41 (2008)
3. Cottrill, C., Pereira, F., Zhao, F., Dias, I., Lim, H., Ben-Akiva, M., Zegras, P.: Future mobility survey: experience in developing a smartphone-based travel survey in Singapore. Transp. Res. Rec.: J. Transp. Res. Board **2354**, 59–67 (2013)

4. Du, J., Aultman-Hall, L.: Increasing the accuracy of trip rate information from passive multi-day gps travel datasets: automatic trip end identification issues. Transp. Res. Part A: Policy Pract. **41**(3), 220–232 (2007)
5. Axhausen, K.W., Schönfelder, S., Wolf, J., Oliveira, M., Samaga, U.: 80 weeks of gps-traces: approaches to enriching the trip information (2003)
6. Schüssler, N., Axhausen, K.W.: Identifying trips and activities and their characteristics from gps raw data without further information (2008)
7. Jun, J., Guensler, R., Ogle, J.: Smoothing methods to minimize impact of global positioning system random error on travel distance, speed, and acceleration profile estimates. Transp. Res. Rec.: J. Transp. Res. Board **1972**, 141–150 (2006)
8. Jariyasunant, J., Sengupta, R., Walker, J.: Overcoming battery life problems of smartphones when creating automated travel diaries. In: Proceedings of the 13th International Conference on Travel Behavior Research (2012)
9. Kumar, S., Paefgen, J., Wilhelm, E., Sarma, S.E.: Integrating on-board diagnostics speed data with sparse gps measurements for vehicle trajectory estimation. In: 2013 Proceedings of SICE Annual Conference (SICE), pp. 2302–2308. IEEE (2013)
10. Tsui, A.W., Lin, W.-C., Chen, W.-J., Huang, P., Chu, H.-H.: Accuracy performance analysis between war driving and war walking in metropolitan wi-fi localization. IEEE Trans. Mobile Comput. **9**(11), 1551–1562 (2010)
11. Stopher, P.R., FitzGerald, C.: Processing gps data from travel surveys
12. Bohte, W., Maat, K.: Deriving and validating trip purposes and travel modes for multi-day gps-based travel surveys: a large-scale application in the Netherlands. Transp. Res. Part C: Emerg. Technol. **17**(3), 285–297 (2009)
13. Schönfelder, S., Axhausen, K.W.: Urban Rhythms and Travel Behaviour: Spatial and Temporal Phenomena of Daily Travel. Ashgate Publishing Ltd. (2010)
14. Parent, C., Spaccapietra, S., Renso, C., Andrienko, G., Andrienko, N., B, V., Damiani, M.L., Gkoulalas-Divanis, A., Macedo, J., Pelekis, N., et al.: Semantic trajectories modeling and analysis. ACM Comput. Surv. (CSUR) **45**(4), 42 (2013)
15. Zhang, N., Kee, J., Loh, G., Tippenhauer, N., Wilhelm, E., Zhou, Y.: Sensg: large-scale deployment of wearable sensors for trip and transport mode logging. Submitted to Transportation Research Board Annual Meeting 2016 (2016)
16. Chang, R., Lee, A., Ghoniem, M., Kosara, R., Ribarsky, W., Yang, J., Suma, E., Ziemkiewicz, C., Kern, D., Sudjianto, A.: Scalable and interactive visual analysis of financial wire transactions for fraud detection. Inf. Vis. **7**(1), 63–76 (2008)
17. Cox, K.C., Eick, S.G., Wills, G.J., Brachman, R.J.: Brief application description; visual data mining: recognizing telephone calling fraud. Data Mining Knowl. Discov. **1**(2), 225–231 (1997)
18. Bostock, M., Ogievetsky, V., Heer, J.: D^3 data-driven documents. IEEE Trans. Vis. Comput. Graph. **17**(12), 2301–2309 (2011)
19. Nisan, N., Roughgarden, T., Tardos, E., Vazirani, V.V.: Algorithmic Game Theory, vol. 1. Cambridge University Press, Cambridge
20. Alon, N., Matias, Y., Szegedy, M.: The space complexity of approximating the frequency moments. In: Proceedings of the Twenty-Eighth Annual ACM Symposium on Theory of Computing, STOC '96, pp. 20–29. ACM, New York (1996)
21. Babcock, B., Babu, S., Datar, M., Motwani, R., Widom, J.: Models and issues in data stream systems. In: Proceedings of the Twenty-first ACM SIGMOD-SIGACT-SIGART Symposium on Principles of Database Systems, PODS '02, pp. 1–16. ACM, New York (2002)
22. Rubinfeld, R., Shapira, A.: Sublinear time algorithms. SIAM J. Discret. Math. **25**(4), 1562–1588 (2011)

Part II
Posters

Modelling Resilience in Interdependent Infrastructures: Real-World Considerations

Mong Soon Sim and Kah Wah Lai

Abstract This presentation discusses the issues that a practitioner may encounter in modelling resilience of a network of interdependent infrastructures. Resilience here focuses on the response and recovery of the infrastructure systems, in the event of disruption. On the model scope, it is not unusual that infrastructures with different operating characteristics have to be taken into account. This is compounded with the challenges to obtain and maintain data for a large network, decipher the specific questions that the user wants to address in the study. On the modelling methodology, traditional OR models are not sufficient to describe the real world applications, as they often do not account for additional requirements. One such requirement is the inclusion of response mechanism of the infrastructures when they encounter disruption at supply or demand side. This presentation will specifically describe eight response mechanisms modelled in a recent study and how they may be incorporated into the traditional OR models as linear constraints via 0-1 indicator and artificial variables. On the implementation, one critical step is to bridge the gap between the modeler and the stakeholder. In order for the model to be used in the users' organization, one strategy is to integrate the model into the user's current Concept of Operations (CONOPS). To do this, the modeler has to look beyond the modeling requirements. One aspect is the model capability to handle an expanding problem size and changing operating environment of the infrastructures. Nowadays, most commercial optimization software allow separation of the data file from the model file; this feature will facilitate this scalability aspect. Lastly, we note that, the key concern is usually not the ability of the optimization algorithm to solve the problem, as commercial optimizers work well for most real-world applications. The difficult issue faced is usually getting the users to use the model well.

M.S. Sim (✉) · K.W. Lai
DSO National Laboratories, 20 Science Park Drive, Singapore 118230, Singapore
e-mail: smongsoo@dso.org.sg

K.W. Lai
e-mail: lkahwah@dso.org.sg

© Springer International Publishing Switzerland 2016
M.-A. Cardin et al. (eds.), *Complex Systems Design & Management Asia*,
Advances in Intelligent Systems and Computing 426,
DOI 10.1007/978-3-319-29643-2_20

267

System Engineering Workbench for Multi-views Systems Methodology with 3DEXPERIENCE Platform. The Aircraft RADAR Use Case

Eliane Fourgeau, Emilio Gomez, Hatim Adli, Christophe Fernandes and Michel Hagege

Abstract The PLM process is developing methodologies to support sustainable production during the whole product lifecycle. The present research makes a focus on the capability to adapt a RFLP framework, integrated in the 3DExperience platform and linked to the Mechatronics disciplines, to the Multi-Views Systems Methodology. This work demonstrates the adaptability of a RFLP framework to a Systems Engineering Methodologies, while taking profit of PLM values for industrial systems production, like an aircraft Radar System. This combination of architecture design and PLM integration increase drastically collaboration increasing the likelihood of detecting product failures early on during its lifecycle validating a virtual product with the customer before detailed design and manufacturing, yielding significant cuts in time-to-market and rework and allowing architects to take the best decision based on global system definition and performance, taking in consideration the disciplines feedbacks Engineering design may be viewed as a decision making process that supports design tradeoffs. The designer makes decisions based on information available and engineering judgment and determines the direction in which the design must proceed. The procedures that need to be adopted, and develops a strategy to perform successive decisions. This interaction shall take in consideration customer feedback. Current Radar Use Case demonstrates the implementation of a Multi-Views methodology, allowing Concurrent Engineering and Integrated Product and Process Development (IPPD)

E. Fourgeau (✉) · E. Gomez · H. Adli · C. Fernandes · M. Hagege
Dassault Systèmes, 10 Rue Marcel Dassault, 78946 Vélizy-Villacoublay, France
e-mail: Eliane.Fourgeau@3ds.com

E. Gomez
e-mail: Emilio.Gomez@3ds.com

C. Fernandes
e-mail: Christophe.Fernandes@3ds.com

M. Hagege
e-mail: Michel.Hagege@3DS.com

© Springer International Publishing Switzerland 2016
M.-A. Cardin et al. (eds.), *Complex Systems Design & Management Asia*,
Advances in Intelligent Systems and Computing 426,
DOI 10.1007/978-3-319-29643-2_21

269

to be performed in parallel and collaboratively. The implementation based on the RFLP framework is extremely helpful to manage complex system design information providing a native traceability and impact analysis capability from the requirements to the operational, functional and component layers of the system, reducing the gap in information. Digital continuity and realistic modeling and interaction between disciplines allow the creation of a digital twin that will be highly valuable during all the lifecycle of the system.

The Methodology to Integrate the Information Concerned About the Safety Between Many Kinds of Materials About the Space System

Nasa Yoshioka, Nobuyuki Kobayashi and Seiko Shirasaka

Abstract To assure that a system is safe, several tasks must be performed, resulting in the compilation of several safety-related documents. It is important that all the information related to safety is developed and maintained consistently. Because of retirements and changes to different department, it is difficult for engineers to pass on their knowledge. To ensure consistency throughout a project lifecycle notwithstanding program-level changes, a method to express the relationship between the safety information described in different documents is needed. Moreover, to address not only specifications but also bases for design formulated by a designer, design knowledge is accumulated together. We aim to improve the understanding of the safety design. In this study, we add two rules to the D-Case in order to achieve three goals. To verify the effectiveness of our method, these rules were applied to a rocket system.

N. Yoshioka (✉) · N. Kobayashi · S. Shirasaka
Keio University, Graduate School of System Design and Management, 4-1-1 Hiyoshi, Kohoku-Ku, Yokohama, Kanagawa, Japan
e-mail: momonunsong3iikanzi@keio.jp

© Springer International Publishing Switzerland 2016
M.-A. Cardin et al. (eds.), *Complex Systems Design & Management Asia*,
Advances in Intelligent Systems and Computing 426,
DOI 10.1007/978-3-319-29643-2_22

Information Architecture for Complex Systems Design and Management

Wael Hafez

Abstract Complex systems are made up of many elements or agents. The behavior and functionalities of complex systems are the result of the interactions amongst the agents making up the system. For man-made systems, the system design is about realizing certain interaction patterns among the system agents that should result in a specific desired behavior or functionality. The system management on the other hand is about ensuring the resilience of such behavior and functionality against erosion and changes affecting the agents and their interaction patterns. That is, system management is achieved by ensuring the consistency and integrity of the designed structural patterns underlying a specific behavior. Systems become complex as the number of agents in the system grows, their variety increases and their interaction patterns overlap. The challenge of complex systems design and management can then be seen to originate mainly from the increasing uncertainties of how changes—either by design or by erosion—propagates across the different structural patterns and ultimately change the system behavior in an unpredictable way. As systems grow more complex, so also the changes they involve, and the unpredictability associated with the overall system behavior. The system management effort and cost thus become extensive and drain the system overall performance and efficiency. The paper argues that managing the communication across the system structural elements (agents) can enable a better management of complex systems uncertainties, and ultimately improve the efficiency of system design and management. The communication management approach is based on using information architecture for capturing the communication (interactions) among the system agents. Integrating system information architecture in the system design and management activities is argued to reduce the uncertainties associated with system operations, increase the resilience of the structural patterns and thus improve the overall system performance and manageability.

W. Hafez (✉)
Independent Researcher, Alexandria, VA, USA
e-mail: w.hafez@semarx.com

© Springer International Publishing Switzerland 2016
M.-A. Cardin et al. (eds.), *Complex Systems Design & Management Asia*,
Advances in Intelligent Systems and Computing 426,
DOI 10.1007/978-3-319-29643-2_23

A CLEAN Way for Sustainable Solid Waste Management System for Singapore

Zhongwang Chua, Jun Wen Fong, Guoquan Lai,
Sheng Yong Kenny Teo, Kevin Williams, Chee Mun Kelvin Wong
and Eng Seng Chia

Abstract From a third world country in the 1960s, Singapore managed to achieve first world status by 2010s within a generation. With the rapid rise in urbanization and economic development, the amount of waste generated by the country had also increased from 1200 tons per person per year in the 1970s to 7200 tons per person per year in 2012. This represented a 2-fold increase in waste generated every decade. In order to keep up with the economic development of Singapore, the Singapore Government understand the need to create a sustainable and efficient Solid-Waste Management (SWM) system for the country. This paper examines the SWM system in Singapore as a Large Scale Engineering System where the goals, stakeholders, boundaries and, complexity were examined. Strategies and alternatives were then proposed to improve the system.

Z. Chua · J.W. Fong · G. Lai · S.Y.K. Teo · K. Williams · C.M.K. Wong · E.S. Chia (✉)
Temasek Defence Systems Institute, National University of Singapore,
Block E1, #05-05, 1 Engineering Drive 2, Singapore S117576, Singapore
e-mail: aaron_chia@nus.edu.sg

© Springer International Publishing Switzerland 2016
M.-A. Cardin et al. (eds.), *Complex Systems Design & Management Asia*,
Advances in Intelligent Systems and Computing 426,
DOI 10.1007/978-3-319-29643-2_24

Mexico Ciudad Segura

Jean Henri Loic Lancelin and Hao Ming Huang

Abstract Mexico Ciudad Segura (Mexico Safe City) project provides an urban security solution to Mexico City, aiming to improve the operational efficiency of police and inter-agency coordination, better communication with citizens, reduce the crime rates and achieve better safety in the city. The Mexico Ciudad Segura solution is a large and complex "System of Systems" (SOS), comprising of a field layer (Cameras, Automatic Plate Number Recognition, Gun Shot Detectors, Emergency Interphones, Mobile communication equipment for the Police forces) connected to 5 Command and Control (C2) Centers and 2 mobile C2s, under the supervision of a Command, Control, Communications, Computers, and Intelligence (C4I) Center. The Mexico Ciudad Segura has been developed by a consortium including Thales and Telmex the Mexican national telecom operator. Three entities of Thales have been involved in the solution design and development. This presentation provides the background information of the project, including the project schedule, engineering dimensioning parameters, concept and overview of operations, share of works and complexity profiler. It reviews the engineering challenges encountered and Thales' response to the challenges by using iterative development, integration and validation lifecycles in all Thales entities. The solution is deployed successfully in Mexico City and key statistics are shown to demonstrate the benefits to the city.

J.H.L. Lancelin (✉) · H.M. Huang
Thales Solutions Asia, 21 Changi North Rise, Singapore 498788, Singapore
e-mail: jean.lancelin@asia.thalesgroup.com

H.M. Huang
e-mail: haoming.huang@asia.thalesgroup.com

© Springer International Publishing Switzerland 2016
M.-A. Cardin et al. (eds.), *Complex Systems Design & Management Asia*,
Advances in Intelligent Systems and Computing 426,
DOI 10.1007/978-3-319-29643-2_25

Product Line Management—Convergence of Architectures and Market Analyses for Operational Benefits in Maritime Safety and Security

Helene Bachatene, Serge Landry, Keng-Hoe Toh and Antoine Truong

Abstract This paper describes a method for product line system architecture applied to the Maritime Safety and Security domain. This approach aims to reconcile interests and concerns from various stakeholders in this domain: (i) End-user organisations and their operational needs, taking into account evolving economic, social and political constraints; (ii) Engineering teams, building product architectures, and managing product and solution development, and (iii) Sales and Marketing Teams, collecting and managing market needs and trends. Architecture and engineering data is exploited, using a customer-centric meta-model implemented to aid decision-makers from both business and technology perspectives, allowing them to understand and predict ROI of development of new system capabilities and to synchronize operational and technical roadmaps for affordable capabilities.

The approach is built using following architecture standards:

ANSI/AIAA G-043A-2012 Guide to preparation of Operational Concept Document

ISO/IEC/IEEE-42010:2011Systems and software engineering—Architecture description

ISO/IEC-42020 (draft) Systems and software engineering—Architecture processes.

H. Bachatene (✉)
Thales Research and Technology, 1 Avenue August Fresnel, 91128 Palaiseau, France
e-mail: helene.bachatene@thalesgroup.com

S. Landry · K.-H. Toh · A. Truong
Thales Solutions ASIA PTE, LTD, 21 Changi North Rise, Singapore 498788, Singapore
e-mail: serge.landry@asia.thalesgroup.com

K.-H. Toh
e-mail: kenghoe.toh@asia.thalesgroup.com

A. Truong
e-mail: antoine.truong@asia.thalesgroup.com

© Springer International Publishing Switzerland 2016
M.-A. Cardin et al. (eds.), *Complex Systems Design & Management Asia,*
Advances in Intelligent Systems and Computing 426,
DOI 10.1007/978-3-319-29643-2_26

Author Index

© Springer International Publishing Switzerland 2016
M.-A. Cardin et al. (eds.), *Complex Systems Design & Management Asia*,
Advances in Intelligent Systems and Computing 426,
DOI 10.1007/978-3-319-29643-2

Printed in the United States
By Bookmasters